Memories of a
Synchronistic Gap Year

Andrew Cole

Memories of a Synchronistic Gap Year

Andrew Cole

Copyright © 2020

ISBN: 978-1-8380486-1-7

Published by Andrew Cole Publishing in conjunction with Writersworld. This book is produced entirely in the UK, is available to order from most book shops in the United Kingdom, and is globally available via UK-based Internet book retailers and www.amazon.com.

Copy edited by Ian Large

Cover design by Jag Lall

Image by Gerd Altmann from Pixabay

www.writersworld.co.uk

WRITERSWORLD
2 Bear Close Flats
Bear Close
Woodstock
Oxfordshire
OX20 1JX
United Kingdom

☎ 01993 812500

☎ +44 1993 812500

The text pages of this book are produced via an independent certification process that ensures the trees from which the paper is produced come from well managed sources that exclude the risk of using illegally logged timber while leaving options to use post-consumer recycled paper as well.

In March 2020 scientists reported that it was possible to translate thoughts into words and sentences, in real time. Or more correctly, to translate the brain activity used in speech into words, sentences and text, then into speech.

This is the first time such a clear revelation that this landmark technology exists and openly revealed in the public arena. Other exciting research continues into this technology including Mind to Machine and Mind to Mind communication, with some major companies investing heavily in this area. All these facts have been widely reported in the media.

However, the author holds that such technologies have been in existence for many years and *Memories of a Synchronistic Gap Year* reveals one such example.

It is a true story of a field trial that took place during the years of 2005 and 2006. Then the technology may have been referred to as Mind Reading or Remote Telepathy but nevertheless it allowed thoughts to be intercepted, interpreted and understood by others.

First written in 2008 but not published for fear of not being believed, it is now published, unaltered and hoped that the reader will understand the book for what it is, an early example of the work, research and testing being done in the field of thought translation. This would of course also be confirmed by the release of any classified documents relating to this trial.

It is a story that spans the globe, Europe, India and Australia and has a strong spiritual element which allows the writer some comfort at the most distressing and traumatic times.

Finally, it offers an insight into how this technology could have been used, rather than for the human good, which is now its likely end purpose.

For the people of Kashmir;

May their prayers be answered and

May their leaders serve them well.

Contents

Southern Spain and Gibraltar Border

The Monday before Easter 2005

I first meet Marian as he was standing outside my flat in La Linea, Southern Spain. He was about 60-years old and carrying six plastic bags in his hands. The morning had been strange in any event.

The previous night Jenny and I had been out drinking till about 4.30 in the morning, which we often did on her days off but what happened that was unusual is that at about 7.30 she had burst into my room and declared that she wanted to drive to Tarifa, a windsurfing paradise about 30km away; Jenny didn't usually get up till the afternoon.

Equally unusual was that I agreed. I don't normally rise until as late as possible. In any event we set out to hire a car and get on our journey. As it was still early no car hire businesses were open, so we adjourned to the diner, a 24-hour bar on the border of Spain and Gibraltar, to wait. Time passed and we both gradually began to get tired and at about 9 o'clock decided that maybe this was not such a great idea after all and elected to return to my flat and sleep some more.

It was as we returned that we first saw him, leant over the railing looking tired and old. We stopped to ask if he was alright and he replied that he wanted a ticket to Poland. My answer was to come in for a cup of tea so we could discuss it but he was insistent that he didn't want tea.

He wanted a ticket to Poland.

He showed us what was inside his carrier bags; they contained some stale, broken, French bread, empty McDonalds cups and used straws, and various other things that could best be described as rubbish. He went on to tell us that he was a science teacher and had come to live in Southern Spain with a friend to teach Physics at the local secondary school but that things did not work out and he was now tired, lonely and wanted to return home to his friends and family in Poland.

He told us his name was Marian Wojtowicz; he had travelled by bus from Poland to Spain and now needed a ticket to return home. As there was a bus station just around the corner from the flat, we decided to take him there and try to get him a ticket home.

This proved difficult because we didn't speak Spanish and the ticket clerk didn't speak English and, further, because it was unclear whether, if put on a bus, Marian was able or in his right mind to look after himself and change buses at the right places. With this we thought it best to buy him some food and something to drink while we considered what to do next.

In the nearby café we ordered some sandwiches, chips and soft drinks and talked some more. He told us that he had been hanging around the bus station for the last four days but no one would help him. He said he had sought help in Gibraltar but that he had been treated worse than a dog. He showed us his ID card and we all chatted happily together.

Strange thing was though he didn't eat the chips, and chips to a hungry man can be a feast.

We agreed to take him to Gibraltar Airport, about a 15-minute walk away and try and get him a ticket to Poland. As we did so Jenny commented that this was like something out of a Quentin Tarantino movie and strangely, as we walked, I looked across at him and he seemed to be smiling quietly to himself.

Something triggered in the back of my mind and I asked myself, how is this going to work?

"The Kingdom of Heaven is spread upon the earth but men do not see it," was my mental reply to myself, a reference to the Gospel of Thomas.

When we arrived at the British Airways desk the booking clerk asked if Marian would be alright to travel alone or if he would need help; his mental state was far from clear, as at times was his understanding of English, so we arranged to have him met at Heathrow and, with this assurance, purchase his ticket before going to the open air lounge and wait for his flight.

We sat and drank tea and talked some more during which time he wrote out a saying in Polish: "Prawdziwych przyjaciół poznaje się w biedzie", which he also translated into English. It read, "A friend in need is a friend indeed".

The flight didn't leave until about seven o'clock in the evening so we talked during the afternoon; Jenny went off to see a friend, while Marian and I enjoyed the sun and the light breeze, chatting during the afternoon about many things, one of which were what countries would he fly over on his way home.

What became clear during this time was that he had a laser-like mind, very sharp and very clever.

As the time neared for his departure I asked a fellow passenger if they would keep an eye on Marian and ensure he made his connecting flight in London. The man, called Joe, said he would be pleased to do so. Joe was returning from his own European travels.

Checking-in time came and as we said goodbye Marian turned to me and said, "Write, you first".

As he left I stood watching, not quite sure of what to make

of the day. Something didn't seem quite right; something had been triggered in my mind. He turned to face me as he went though the departure gate and we looked, maybe both a little unsure of each other.

It had been an odd day and as I considered what had happened I felt sure that I had read something about this man somewhere.

It was after a night's sleep that I started to further recall something that I had read in one of the quality Sunday newspapers. As I thought about it more and more it became clearer and clearer. Indeed I had read that, "Marian was at it again".

Apparently, someone called Marian had in the past travelled the world and tried to get people to buy him a ticket home.

The article had said that if you meet Marian, be careful what you say because sometimes he gave people ten million pounds and sometimes just one million pounds.

The author of the article assumed that this was because Marian was getting distrusting in his old age. The more I thought about it the more I recalled of the article. So much so that I remembered that when I first read it somehow I knew I would meet him, somehow instinctively I knew I would.

It reminded me of someone who had worked in Gibraltar and had known they would win the lottery, which they did; somewhere inside them was the knowledge that they would one day win the lottery, instinctive knowledge.

In was this idea of instinctive knowledge; this idea of innate understanding, that was to take up much of my time and thoughts over the next few months.

The Next Three Months

In the days that followed I spent time considering what I was going to do with my reward, checked my bank account to see if any money had been paid in and generally thought a lot about the whole experience.

I considered writing to Marian as he had instructed but resisted for a while, until one morning I woke up and decided to send him a hand-written letter.

It was while doing this that I considered his question about which countries he would fly across. It then occurred to me that we could have driven to Malaga Airport, about an hour away and from there he could have flown directly to Poland rather than flying from Gibraltar to Heathrow and then onto Poland; the direct route would have been a lot quicker and a lot cheaper.

It was only then I fully understood; that was what he had really been suggesting when he asked the question, about which countries he would fly across.

Time passed, no reward was forthcoming but I was still intrigued. How did I know I would meet him? This raised questions for me about why he was at my flat, at that time; was he there to meet me or was it just a coincidence? Had I met him before? Indeed, at times, I questioned if I had ever read the article at all.

It was from this point that I started to experience a whole array of... Just coincidences.

For example, I would leave the flat with the intention of spontaneously going to see someone at their home, only to bump into them walking down the street, or I would be sitting in the flat considering what to do and a friend would ring to arrange an impromptu night out, at exactly that moment.

My whole life was becoming a whole range of wonderful... Just coincidences, to such an extent that I decided to revisit some past study on the subject.

* * *

For this the starting point was Carl Jung's work. Jung, a psychologist, had developed the idea of synchronicity, which can easily be understood as meaningful coincidences with no apparent causal effect.

Jung, who was a Gnostic, held that the religious or spiritual nature of humankind dwelt within, as opposed to the view that religion was something external.

For him, religiousness, spirituality, a greater force or whichever name we choose to use was a living essence that lived within the human psyche, was part of the human psychic.

This principle underpinned the two innovative concepts that he created.

His second concept was that of the collective subconscious; this held that matter connects us all, our bodies in essence being no more than the total sum of what we had consumed during our lifetime, all our food both coming from and returning to the planet.

He argued that the principle worked in exactly the same way with our psychic or our subconscious, which he believed to be connected at a deep level.

His theory held that, beyond our ideas of our own personal identity or self, lay a second layer of consciousness in which was carried innate knowledge such as primeval fears, and beyond this level of consciousness lay a further level, the subconscious by which we are all connected. It was at this level of consciousness that synchronicities took place. It was at this level of consciousness that we could know things.

My studies included the Bible, the Koran, and the Bhagavad Gita, a book on Tibet and Buddhism plus an essay found on the internet called *The Depths of Existence* by Mihai Drăgănescu, who was head of the Romanian Academy for Artificial Intelligence. This is a fascinating essay although there was much of it that I did not fully understand. It broadly outlined the differences and similarities between the human mind and machines. The main difference, Drăgănescu sees as being the spiritual nature of humankind.

Further, he highlights that in the field of Quantum Mechanics, thinking can cause matter to act upon itself, and thus thinking can affect matter.

This would seem to be a massive scientific point!

* * *

It was a time of reading, thinking, drinking and listening to music, being with friends and family in a state of freedom; in short, living the life. I would spend hours contemplating the clouds, seeing different patterns or faces in them and just enjoying being, seeing the interconnectedness of nature and seeing the human spirit that dances between individuals.

Many synchronicities or coincidences happened during this time. For example, I could not shake the thought of the newspaper article about Marian and the reward. One day, while considering it and flicking though the Bhagavad Gita, it opened on a sutra that encourages giving up expectation of personal rewards as a way of more clearly seeing the truth.

Also, my youngest daughter Helen came to visit and we decided to go to the local beach in La Linea. While there she showed me her iPod which she had recently bought and I asked if I could use it. The first tune that played was *Have Yourself a Merry Little Christmas*, a tune that had special significance for us both, as it was a song she had sang as a young girl in primary school and was one of those special songs that was part of the history between us. Out of over 4,000 tunes on the iPod it was the first one played; she hadn't known it was there. These are just two of many possible examples.

It was also during this visit that she asked me what I wanted to do, as I had reached a crossroads in my life. After thinking I told her I wanted to travel, to go maybe to India or Australia or Rome, all for different spiritual reasons; India because of its rich spiritual history, Australia to do some research into the spirituality of the indigenous Aboriginal or Rome to work as a kitchen hand in the Vatican.

She encouraged me to do something, to do some or all of the above.

After Helen had returned to England I continued in much the same vein as before, developing different thoughts and ideas on Theology and Science.

It was to my mind possible to work towards a scientific understanding of God. (By God, I mean Allah, Krishna, Yahweh, Jehovah, that which is ultimate reality or whichever name we chose to describe this Greater Force.)

For me there was no doubt that a greater power existed and interacted with the world; something that could be understood by examining any or all of the world's great religions but, more importantly, something that could be experienced personally.

* * *

There is a bar in Gibraltar called the Cannon Bar, situated just behind the cathedral, where I had met the retired Bishop of Gibraltar. He was about 80-years old and sometimes used the bar for his post-dinner brandy and coffee. He had told me the story of his becoming a priest from his early days in Ireland and how he ended up in Gibraltar via his own personal story, via his own set of personal circumstances.

Once, when I met him, he had a very badly bruised left ear, so I asked what had happened. He told me that one night he felt a great force trying to push him out of his bed, so forceful that he couldn't resist it no matter how hard he tried. The force had eventually succeeded in pushing him out of his bed and he had banged his ear on the bedside table. This story intrigued me because here was a senior member of the Church suggesting and accepting that forces outside of what is generally believed to exist, do exist.

On this occasion I had gone to the Cannon Bar specifically to see him because there were some points I wanted to discuss regarding my studies and travels. After buying a pint of beer I sat outside hoping he would turn up. It was while sitting outside and staring into a vacant space that it occurred to me that a small bush just opposite had the shape of a face and as I thought about my plans and my questions it seemed as if the bush would sway in the light breeze either confirming or rejecting different thoughts.

The imaged face of the bush nodded either from side to side as a no, or up and down as a yes. I must have been engaged in this imaginary conversation for quite some time, as the barman came over and said it was the first time he had known a pint of beer to aid a trance-like state. I smiled and left feeling that I had the answers to the questions that I wanted.

The plan had been set; it was to go to India. Initially, this was to have been in August with a friend but situation and circumstances dictated that I would leave Spain in the June and travel to India alone but first I needed to return to England to renew my passport, get a visa, have the inoculations and visit family and friends.

Back in England

My dad's flat would provide a base for me while in England. He lived in warden accommodation in Thame, Oxfordshire with a spare room that could be let to guests. From here I would be able to firm up the details of my travels, get to London to arrange the necessary details, visit friends in Oxford and my two daughters, Helen in Bushey and Michelle in Cambridge.

The plan was to fly to Delhi before heading to Leh and Ladakh in Kashmir with a view to getting to Matho Gompa, a monastery in the Himalayas, where the monks would go into meditation for two months each year before performing some unbelievable feats at the Ladakh Festival. But first I needed to sort out the details and paperwork for the journey.

One of my first trips, after arriving back in England, was to Oxford to visit my friends Gill and Stuart. Taking the bus but not knowing where to get off I asked two elderly ladies. Luckily, I asked them at exactly the stop I needed to get off at. It seemed to me that the coincidences were still working.

I was to stay with my friends for a few days before checking into Nanford Guest House in Oxford. Strangely, the guesthouse, chosen at random, had pictures of the Himalayas all around the walls. The owners were frequent visitors to the area and were able to offer me encouragement and advice for my forthcoming journey.

Of course I took this as a route confirmation sign; meaning that I was convinced that I was still on my right path.

While in Oxford I also visited my old university, Westminster College. I had taken a mid-life degree in

Theology there, in the early nineties, and now wanted to revisit the Alistair Hardy Centre which used to conduct research into personal religious experiences, but because the university had been taken over by the ex-poly Brookes, the centre had moved and no one knew where it had relocated to.

That evening I went to a pub on the Iffley Road and whilst out met a man who was starting a new teaching job at a nearby school in Didcot. We discussed many things but he told me a story, which I remember clearly, about a friend of his called David Bailey. His friend David had at one time been a very wealthy man whose life had been ripped apart by the loss of his family in a traffic accident. Devastated by his loss he had given away all his wealth and gone to live in a monastery in Tibet. It was a very sad story but for some reason I felt his name would be important and that it needed to be remembered.

The other thing that I wanted to do while in Oxford was visit Blackwell bookshop; the shop was situated just across the road from the Bodleian Library and Sheldonian Theatre. A small fronted shop that opened up as you entered it, into a vast store with three floors that housed a wonderful and extensive range of books. After browsing for a time I came across a book by Alister McGrath called *Dawkins' God*.

McGrath was Professor of Historical Theology at the University of Oxford and challenged Richard Dawkins' views on God. The criticism is centred on the dispute between evolution and design. This was a theme that was to follow me to my next destination, which would be Cambridge.

* * *

On my last night in Oxford, Gill and Stuart threw me a party, during which they gave me a card, in which they had written, "We know wherever you go you will be alright because God will be with you."

It was a sentiment I was to rely on heavily in the not too distant future.

The following morning they took me to the station, where I brought a copy of *The New Scientist* because the headlines read, 'Either Free Will or Reality is An Illusion'.

The article referred to the work being done on protons in Geneva and the Quantum Theory of Entanglement, which has discovered that protons, many miles apart, react simultaneously. I read, enjoyed and thought about the discovery on my journey to Cambridge.

* * *

My daughter Michelle was just finishing her studies at Homerton College and wanted help moving back home to Bushey. During my stay she spent time completing her studies, saying goodbye to friends and colleagues and packing while I spent some time visiting Cambridge. She would drop me off at King's College so I could walk though the beautiful grounds to the town centre. The town is a busy bustling city and on a sunny Saturday as I walked though it I was handed a leaflet; it was a flyer for a lecture on Intelligent Design vs. Evolution.

Of course I wanted to go to the lecture and spent the time waiting for it to start listening to the street buskers, some of whom were playing classical music and in particular a tenor with an outstanding voice. I also visited the local Border book shop, where, as I bent down to tie my shoe lace, found a book called *365 Tao*, which had a verse for each day of the year.

The first page it opened on was verse number 269, which related to a parable about tying shoe laces, for me a definite sign, which made me smile and with that I bought the book and went to the lecture.

The topic of the lecture is one that is becoming increasingly important in today's secular society. After enjoying the lecture I returned to Michelle's house by bus, travelling on the top deck and meeting some interesting young girls who shared their vodka and coke, in a coke bottle, and their small spiff with me.

During that evening Michelle and I enjoyed each other's company and discussed what had happened to us during our different days. It was at that point that it dawned on me that Tao, 269, related to 26th September, which is also Michelle's birthday and with that I knew that the book must be for her. It was also the only Tao number that matched in this way.

It was during that evening that Jenny rang, from Gibraltar, to say I had received a letter from Marian.

Excitedly I asked her to open it and read it. The gist of the letter was that after returning to Poland, Marian was unhappy and wanted to come and live with me in Spain. With this I had to reconsider, maybe I was wrong about the article, maybe he was just a poor Polish man trying to find his way home, maybe I had imagined the article in the paper, maybe there would be no reward.

Suppressing my disappointment I decided to send him a proper type-written letter. To do this the next day I went to an internet café near the town centre. It was run by a Jordanian man with maps and pictures of his homeland all about the place. There was a back room where friends shared hubble-bubble pipes; the whole atmosphere was really warm and friendly.

After writing to Marian I checked my e-mails and found that Joe had written to me. This was odd I thought because I had met them both in Gibraltar some time ago and now they were both getting in contact at the same time. Joe and I exchanged e-mails and arranged to meet in London in the near future.

With that I sat down to have a cup of tea and got into conversation with a friend of the owner who was doing a PhD in Psychology at Homerton College. He was critical of Carl Jung's work, on the grounds that Jung was a Communist, which I thought was both very narrow minded and non-academic for someone doing a PhD.

The following day Michelle and I packed the last of her things and as we did so she told me that the clock on her car kept skipping forwards and backwards an hour each time she drove down her road. Without giving it another thought we drove to Bushey.

She was to move back to live with her mother and her grandmother; her mother Lin had left me some years ago to live with her own mother Pat, taking the children with her. Over the years we had all retained relationships of varying degrees. Because of Michelle's move and my planned travels I would also spend a few nights staying there.

My friends Gill and Stuart from Oxford knew a local Baptist minister, whose parish was nearby in South Oxhey and, because of the type of journey that I was on, I decided to go to visit her to discuss some theological ideas. I told the minister and her friend of my travel plans and, by coincidence, it turned out that they had themselves planned to travel to Ladakh in Kashmir but never got there for some reason.

In addition they had both studied under one of the same lecturers as me, Peggy Morgan, who after leaving Westminster College had gone on to lecture at Maudlin College, Oxford. They were also able to tell me that the Alistair Hardy Research Centre had been relocated to Lampiter University in Wales, via Peggy Morgan's garage.

We shared a number of similar theological ideas, which included the importance of psychology and of Carl Jung's work within that field.

There was a book on the sofa about India, religion and artificial intelligence, which I noted. It seemed to me that there were a number of synchronistic events in our meeting.

By this time Helen, my youngest daughter, had gone to work in Canada. Michelle was spending time with friends and Lin was looking after her mother. For me it was time to go to London to collect a new passport, get the visa, have the inoculations and buy some equipment and the plane ticket, before travelling to India.

Off To London

As I planned to leave England in about two weeks and there was still a lot to be done my plan was to stay in London for about week then return to Thame, say goodbye to my dad, then back to London before flying to Delhi on 12th July. King's Cross was chosen as a location because it was central, had easy access, offered inexpensive hotels and was where I was born. I checked in at a small hotel just across the road from the main train station.

My first day would be spent going to the passport office in Victoria to collect a new passport. I was comfortable in London, although I hadn't been there for a long time as I had being living in Spain for seven years but it was where I was born and I considered myself a Londoner.

Whilst in Victoria I decided to visit another old college of mine, also called Westminster but this one was the catering college. I had trained there in my late teens as a chef, many years ago. Certain memories came flooding back as I walked towards Vincent Square, then back toward Victoria Station and in doing so I came across a small shop with religious images in the window, so I decided to go inside.

As I looked around the shop there was a lady with a string of rosary beads in her hand and strange chanting noises were coming from a room below. I asked her what was happening and she told me that there was a meeting taking place and that this was a prayer group of The Church of Padre Pio, an unorthodox part of the Catholic Church. She explained that the chanting was the congregation saying their Hail Marys and with that she invited me to join them in their worship, which I accepted.

I had heard of Padre Pio, as he had recently been canonized, thus becoming a saint but other than that knew little of him. The meeting room was small but full and held about 40 people who would come and go during the service and after about half an hour I had to go myself to keep an appointment with the passport office.

The next day was a trip to India House to collect the visa, and then the day after a trip to Covent Garden to buy some boots, rucksack, mosquito net, torch, knife etc. and some warm clothing for the Himalayas. Then on to Piccadilly to confirm and collect my plane ticket from British Airways. Everything was falling into place very nicely but I was still uncomfortable about the Marian situation and of course the reward.

It was difficult to accept that I had imagined the article and that he was just a poor Polish man trying to find his way home.

The only thing I could think to do was to take advantage of being in London and go to a library and search the newspaper archives of past copies of the Sunday papers in the hope of finding the article and thus putting my mind at rest. So the following day that's what I did.

The first library I went to was just off Leicester Square but after some hours of research and loads of help from the librarian I was still unable to find any references either using back copies of the Sunday papers or internet searches. As a last resort the librarian told me that all she could suggest was to try the library in Swiss Cottage because they held back copies for longer than the Leicester Square branch. So that's what I did.

The Swiss Cottage library is great, with many interesting quotes from a range of famous people around the side of the building. Again, with a lot of help from the librarian, my search was unsuccessful. The nearest I came was a copy of *The Independent on Sunday* which had some pages missing where I thought the article may have been. Also, as I walked down the steps to leave the library, a bus drove past with an advert that read:

'The Independent, Take a Fresh Look'

Did that mean anything? I didn't know but I was meeting Joe tomorrow and maybe I would get some clues from him.

Joe was waiting for me outside King's Cross Station; he had with him his rucksack because he was on his way to protest at the G8 Summit in Edinburgh. We decided to adjourn to a nearby pub to catch up. He told me that after they left Gibraltar he had ensured Marian was in capable hands at Heathrow before he left him but other than that he couldn't tell me anything else.

He did tell me that he had developed his own website and written about his own travel experiences; it was called Whatpai, which means atonement in Maori. Somewhat disappointed that I had been unable to find out any more details about the Marian situation we said goodbye. With that I decided to push the whole Marian experience to the back of my mind and concentrate on the journey ahead, consoling myself with some synchronistic events that Joe had told me about.

The next few days in London were busy with lots of different and interesting things happening. First there was a demonstration by people of the Congo, protesting at the abuse of their countrymen and the misuse of their country's natural resources. Then, on the Saturday during the day, there was a Gay Pride parade in Trafalgar Square and in the afternoon and evening Live 8 was taking place in Hyde Park;

all of these events have their own small side stories.

On the Tuesday I returned to Thame to spend a few days and say goodbye to my dad.

On the Wednesday London won the bid for the 2012 Olympics, as the local school held its annual sports day.

On the Thursday morning my dad woke me to tell me, "The bastards have bombed King's Cross."

It was the following Monday that I returned to London, the street of the hotel I had been staying in was now packed with news teams from around the world, reporting on the terrorist attack. The buses were full on the lower decks but empty upstairs, the underground was very quiet and there was generally a very sombre atmosphere everywhere.

I got into conversation with some office workers, their first day back at work since the bombings; they retold their stories from the day of the attack. I bought some flowers to lay at the church opposite Euston Station, bought some travel insurance, gave away some REM tickets because the gig had been put back a week because of the bombings and prepared to leave for India the next day.

The Indian Experience

You can get to meet some interesting people on long haul flights. For example, there was the woman who was going to meet her husband and then head off to Dharamsala to hear the Dalai Lama speak and the young man who was heading to Calcutta to work for a week in an orphanage before going on to Goa. With conversations, meals, drinking and sleeping the flight soon passed and we arrived in Delhi about midnight local time.

It was hot, humid and lively and as I had pre-booked the Centaur Hotel near the airport I just needed a local taxi to take me there. From the many tourist touts I chose one to help, his name was Dinesh. He readily arranged the transport and said he would ring me next day to give me a local tour of the city.

True to his word, first thing next morning Dinesh rang. He organized a taxi tour of some of the highlights of Delhi such as India Gate, Humayun's Tomb, The Baha'i Temple and Qutb Minar, and the Gandhi Memorial. It was a great day, seeing and experiencing sights, sounds, tastes and smells I had never experienced before. These also included both a Hindu funeral and a Hindu wedding, both very different to anything I had ever witnessed.

Also I liked the billboards with slogans such as 'Two things you should always follow; your dreams and traffic signs'.

As the day drew to an end he took me back to his office and it was there after enjoying a curry with him and his colleagues that we planned the next leg of my trip. Because he was unable to book a flight for me to Leh, in Kashmir, he instead booked a flight to Srinagar, also in Kashmir but further west. We agreed that next morning he would collect me from my hotel and put me on the plane. It had been an excellent day rich in new experiences.

Although an internal flight the security for the flight to Srinagar was tight and as we landed I noticed that down each side of the runway were army tents. As we disembarked the captain spoke over the intercom thanking us for flying with Indian Airways and that he and his co-pilot David Bailey hoped we had enjoyed our flight.

The name David Bailey immediately registered as the name of the man who lost his family and went off to Tibet to become a monk. Although not thinking for a moment it was the same man, it was the name I had known would be important when I had first heard his story in Oxford.

Taking this coincidence as an absolute route confirmation sign, meaning that I totally believed that I was where I should be and absolutely on my right path, I delighted in my new surroundings. Knowing that I was where I should be, although I didn't know why.

After clearing arrivals I was met by my host in Kashmir, Ashley. He had a broad smile and sparkling eyes as he packed my rucksack into the back of his 4 x 4. We drove for less than an hour before arriving at a lake and then boarded a small boat to paddle to a larger moored houseboat. It was Lake Dal and although in a totally strange environment with total strangers, far away from that which was familiar, I felt totally at peace.

The houseboat was handcrafted from cedar and pinewood and sat gently upon the lake. On arrival I was served with fresh, locally picked Kashmiri tea, coconut macaroons and fresh mangos and as I sat on the deck of the houseboat, overlooking the lake, watching the eagles soaring in the wind, in the foothills of the Himalayas, I felt a certain bliss. And decided to stay for a while.

Ashley joined me in the early evening and as the call to prayer echoed across the lake I felt a real sense of belonging, a real sense that something inside of me had been deeply touched. It was truly beautiful.

My first full day was spent relaxing on the houseboat learning a little of the history of Kashmir. It had many centuries ago been a country in its own right with its own monarchy. Since partitioning in 1947 it is now split in two, part in India and part in Pakistan; it is bordered by Afghanistan and Tibet, China.

More recently it had suffered from internal strife as the front line in a war between India and Pakistan. This had led to many thousands of innocent local people being killed. The war dates back to the rigged election of 1989 and continues to this day, hence the heavy army presence at the airport.

The lake was a busy place with many small, colourful boats that I learnt were called shikaras, operating as floating shops or taxis to ferry people to the shoreline. It was set in a valley and hence surrounded by gently rolling hills, one of which was called The Throne of Solomon. It was over this hill that the sun would set, giving wonderful and inspiring views.

Also, during that first day, Ashley and I got into different conversations. We spoke of God, Allah for Ashley, and agreed that just as I would say water, he would say pani; but that we were both talking about the same thing but only using different language.

He inquired of the recent bombing in London, for it had only been a few days before and with that we also got into conversation regarding terrorism. I told him my view was that one man's terrorist is another man's freedom fighter. With that we got into conversation regarding the current situation in Kashmir and how some would call those involved in the local fighting terrorists, others may call them separatists, some may call them insurgents and still others may call them freedom fighters, depending on different perspectives. It was an interesting, educational and thought-provoking day.

And the longer I sat on the deck of the houseboat with my Kashmiri tea, watching the eagles fly, feeding the friendly local birds, watching the sunset over the Throne of Solomon, listening to the call to prayer, the more I loved it.

On my second full day I met some of Ashley's friends and arranged a programme of local tours to do during my stay. We had agreed that I would stay for a week to start with and tours would be arranged for me to visit some of the local mosques, the Mughal Gardens, and some homes of his friends. The highlight would be a visit to the Sufi Baba, an Islamic holy man, which I was really looking forward to. I also met Abdul, a man about my age and someone I would form a strong personal bond with.

After the details had been arranged Ashley and I took a shikara to the shoreline so he could show me some of the town. This was an experience in itself; we saw a policeman trying to score some hashish from a local man and Ashley told me that everybody smokes hash there. I gave a 100 Rupee note to a small pretty young beggar girl with her mother, to which the child gave such a scream of delight, as if she had won the lottery. It was worth about one English pound. Ashley also wanted me to meet a friend of his who owned a café so we went there to have some tea.

This café owner, Aziz, was delighted to see his friend.

Ashley was obviously a well-liked and well-respected man, as everyone we met was pleased to see and greet him. Aziz joined us for tea, sandwiches and conversation. He told me more about the history of Kashmir and The Last Viceroy of India, we discussed the Kashmiri borders, and something called the Dixon Plan, a diplomatic initiative to resolve the local problems, in addition to terrorism, freedom and beliefs.

He also spoke of Palestine and the need to resolve conflict there. During the conversation he also told me how Osama Bin Laden had received a hero's welcome in nearby Afghanistan.

At that point I asked a question as to who was the greater man, Hitler or Gandhi; Aziz's reply was Gandhi. This was just about the end of our conversation.

It had been a good conversation and a very enjoyable day with no hint of what was to follow.

Thursday Night

It was tonight that it started.

I had gone to bed in a happy relaxed state of mind. The wind had gently moved the net curtains in my room as I said my hellos or my prayers, to my God. But.

I was woken up in the early hours of the morning by a loud noise and the chanting of, "Hitler, Hitler, Hitler."

Repeated over and over and over again.

"Hitler, Hitler, Hitler."

I was frightened, alarmed, and concerned; this chanting lasted for some time.

The sounds seemed to be coming from a loudspeaker

somewhere nearby and I was sure the chanting referred to the conversation that I had earlier in the day with Ashley and Aziz in the café.

After a short pause the chanting started again, this time with, "There's somebody in Gibraltar, there's somebody in Gibraltar."

Again, repeated over and over.

I was starting to get very frightened, indeed there was somebody in Gibraltar, for I had many friends there but how did anyone know I had been in Gibraltar. I struggled to understand what was happening.

Then, after a short pause, "There's somebody in Gibraltar, there's somebody in Gibraltar, we're going to kill your daughter, we're going to kill your daughter, there's somebody in Gibraltar, and we're going to kill your daughter."

This took my stress levels and fear levels to new heights because now it seemed that my family was being threatened.

It was about now that I really started to sweat, really started to worry, and really started to be concerned.

Then I heard the sounds of tanks rolling along the street, then the sound of marching troops, the sounds of gunfire. All of this was too much; I got out of bed in a state of fear and confusion and hid in the toilet. My heart was thumping and my pulse racing yet my mind was clear. I was experiencing levels of fear and tension I had never experienced before.

The sounds continued, yet with a clear mind I thought, what was the point of hiding in the toilet? It seemed that my body was reacting in a physiological way but my mind was reacting in a logical way. Gradually my mind started to calm the physiological reactions of my body and I decided to return to bed in an attempt to conquer the fear.

Eventually the chanting and the sounds faded and were replaced with other unfamiliar sounds but these were sounds of dawn breaking in a strange environment, the sound of crows landing on the tin roof of the houseboat, the sound of birds waking, birds that I had never heard before. Light started to filter through the window; I started to feel safer and calmer and must have eventually gently dropped off to sleep.

Friday

The next morning there was a knock at my door to tell me my breakfast was ready; everything on the houseboat seemed calm and normal, which in itself calmed me. As I sat and ate my breakfast of local honey, toast and Kashmiri tea I decided to keep the events of the previous night to myself. The plan for the day was to visit some of the local mosques and shrines with Abdul.

We took a shikara to the shoreline and walked towards Sultan Arfaen's Shrine, which was to be our first stop. As we walked and talked it became clear that Abdul would be great company, his easy manner, sense of humour and open expression were just a few of the things I liked about him.

He showed me places of interest along our route and we spoke about our families; he had recently lost his wife of many years. I told him about my ex-wife Lin and how she had left me many years ago. His reply was that it was still my duty to make sure she was alright and look after her, the best I could; this was a man's duty to his family, a sentiment I really liked. We walked though the city gates and started the climb to the shrine.

Outside he treated me to some corn pellets so I could feed the pigeons and explained about the pre-worship washing areas before we then entered the shrine.

The outer area was full of beggars of various ages from young to old, all with their hands painted with henna; it was in this area we gained permission to enter, it was also where we took off our sandals. There were a few steps as we entered the shrine proper.

Inside, the shrine had a sense of great peace, it was busy with some people just sitting, others reading the Koran, all were men but the women were allowed to watch their husbands reading and studying via an open window, which I thought was wonderful; women lovingly watching their men as they spoke lovingly to their God.

We sat with our back to a wall and just looked; I was mesmerized by the calligraphy. After about 15 minutes Abdul asked if I wanted to leave but I didn't want to go, I just wanted to stay and be lost in my own thoughts and the peacefulness of it all.

Eventually, recalling the events of the previous night I decided to say a small prayer: "If it be your will."

With that a small breeze, maybe from an overhead fan, brushed my face, making me feel deeply happy and I said hello to my God.

As we left we were handed some Holy sweets, which in turn we gave to the beggars. It had been great, I had a really, really special time there and Abdul and I walked down the steps like two schoolboys together.

The next stop would be Shah Hamden Mosque. We walked through the old town, a beautiful area rich in architecture. Along the way we passed some army posts that were camouflaged into the streets.

We passed the tomb of Isa, the Koran's name for Jesus. Some believe that the body of Jesus was laid to rest here; in fact there are many books and TV documentaries on the subject but at this moment it was closed.

As we approached the mosque, Abdul's mischievous side showed itself. He told me to tell the six clerics sitting outside the mosque that I was from the British press because then they would make a big fuss of us. I duly played along, which was fun. The building was almost empty inside and highly decorated; the outside was totally made from wood and looked more Buddhist than Islamist, and was very impressive. Although I wasn't allowed inside because I was not a Muslim I could enjoy it through its open windows.

From here it was on to Jama Masjid Mosque, which was enormous and could hold many thousands of people; it had been taken over by the terrorists some years ago, at the height of the troubles but was now once again a busy, working, peaceful mosque. It was impressive both in its size and the fact that each of the many pillars supporting the roof is made from a single tree.

We walked on again, this time passing shops and markets. Abdul told me many of these businesses had been closed at the height of the troubles and curfews imposed. Today, like much of Kashmir, they are trying to re-establish themselves. As we walked we met different people and everyone seemed so happy and open. Very warm and welcoming people, the Kashmiris.

Our last visit was to the tomb of Zain-ul-Adidin, the burial place of Kashmir's former monarchy. From here we were off to the town centre for some refreshments and to look at the local bookshops. It had been both a very full and very enjoyable day.

That evening some of Ashley's friends came to the houseboat and we talked for some time after dinner. I was told that at the height of the troubles, terrorists would come in the middle of the night, to people's homes and demand to be hidden, using their guns to threaten the householders. Then, they would take the mother or the prettiest daughter and rape them, in front of the family.

Somebody suggested that it was because the Russians retreated from Afghanistan that the Taliban were able to come across the mountains and bring terror to Kashmir in the late eighties.

We discussed the possibility of nuclear war. It was suggested that if there were a nuclear war it could be in Kashmir; both India and Pakistan have nuclear weapons and the distance between them is short, thus allowing for a swift attack. Someone else said that many of those who can remember think it was better and safer under British rule, while Ashley made the point that Kashmir was now no more dangerous than any other part of India.

What also became clear was they were all very proud to be Kashmiri and that the average Kashmiri had suffered greatly because of the past troubles, not least because of the enormous loss of innocent local lives.

After they had gone I got the quarter bottle of whiskey I had bought earlier from the floating shop, noted that I was now the only tourist staying on the houseboat and hoped the whiskey would do its job and help me sleep soundly. Which it did.

Saturday

After a good night's sleep I felt refreshed and recovered. Today's trip would be a shikara journey across Lake Dal to Golden Island and then on to the Mughal Gardens. Ahmed was to be my guide today; he was short with greying hair and beard, with big sparkly eyes, always wore traditional dress including hat, loved singing, spoke great English and proudly told me he was a Shia Muslim.

He, plus our wallah and I, set off past the floating shops, past the floating police station toward Golden Island. I felt very spoilt to have both a guide and a wallah all to myself. Ashley's wife had given us a packed lunch because we would be out for the day.

The day started calmly enough as we paddled across the lake. Golden Island was fine, not too much to write about but I did see possibly the most beautiful woman I had ever seen in my life there, her eyes were as large as saucers and so clear and open you could almost see her soul.

From there we continued across the lake enjoying the views of the surrounding hills, onward to the Mughal Gardens. We walked the gardens for some time before stopping for some tea.

It was while sitting in the café of the garden that two lovely little girls came up to us and pointed at me and said, "Is that a kuffer?"

They were about six years old. It seemed they lived in the mountains and had never seen a westerner before.

As Ahmed and I sat relaxing in the outdoor café it almost seemed that I had been there before, something was familiar, almost in a dream-like state. Maybe I was experiencing some type of déjà vu or maybe it was a place I had dreamed of. In any event I felt very comfortable and relaxed and at home. We finished our tea and set off back to the shikara to have our lunch by the lakeside but as we were doing so, a massive convey of army trucks drove by.

There were so many trucks, some empty but some full of soldiers, all had painted faces and looked like they were ready for combat. There was one soldier who stayed in my mind's eye, he looked so keen, he looked like he was ready to kill or be killed, standing tall and proud in the back of one of the trucks with his neckpiece flapping in the wind and his blacked out face. I had never seen men going to fight before, to kill or be killed and found the whole experience both frightening and exciting. Ahmed seemed quite unmoved by the whole situation but suggested maybe it would be better to have our lunch on the lake.

We rowed out to the middle of the lake and as we did so the firing started. I started to get frightened, never having experienced live gunfire before. Ahmed told me not to worry and that it was just a wedding celebration. He and our wallah both decided to take a nap.

Me, I was extremely scared, well outside of my comfort zone, I knew it wasn't a wedding party because of what I had just seen and by the sounds coming from the hillside.

I decided to pray, "Yet though I walk in the valley of death, I fear no evil for thou art with me rod and staff."

And I guess that Kashmir had, in the past, been a valley of death and thought the prayer very appropriate.

Still scared and frightened I watched as a small cloud floated over the area where the firing was coming from. Then I looked down to the lake and there on top of a floating lily leaf, two twigs had formed the sign of a cross; it was there for a moment and then it was gone. I said thank you to my God, for my sign, and started to relax a little more.

What I also noticed was that all of the boats on the lake were stationary. Did that mean everyone knew what was going to happen or was that just standard practice in these situations? My guide and my wallah didn't want to take any questions on the subject and with that I understood that, in these situations, you don't ask questions.

In any event I was extremely grateful for my sign. The firing lasted for about an hour, after which we slowly made our way back to the houseboat.

That evening as the sun set, the call to prayer was followed by a sound of great wailing and great sorrow. I understood this to mean that the terrorists had been defeated. But I started to get a little confused when the wailing was replaced with the song, 'All we are saying is give peace a chance', followed by, 'Dancing around the Christmas tree won't you come and play with me.'

Then, 'Let me love you in paradise.'

Then back to, 'Give peace a chance.'

I reasoned that differing sounds and songs represented differing opinions and that whoever controlled the microphone in the mosque could also exert great influence, if indeed the sounds were coming from the loudspeakers of the mosque.

I started to feel even more uneasy as the manager of the houseboat came round and started closing the wooden window shutters. This was something he had never done before. In view of the day's events and my increasing levels of tension, anxiety and uncertainty I decided to have an early night.

The shutters on my windows weren't closed so I tried to

close them but they didn't work. This made me feel even more vulnerable. I went to bed and tried to sleep but this proved difficult. The events of a few nights ago came flooding back and I started to sweat. Now each sound increased my tension levels; it seemed that the sounds were coming from the ceiling fan, the hum and rhythm of which made me even more nervous. I tossed and turned in my bed, which now seemed too small.

Throwing the covers off then pulling them back on again, increasingly fretful, increasingly concerned, my mind started racing. I then started to hear repetitive sounds, sounds that matched with so many different words. Rhythmic sounds in tune with my heartbeat and as these sounds increased in frequency, so the beat of my heart increased and followed.

Then I heard the bloodcurdling sound of a man being killed followed by a large cat meowing loudly, as if to mask his screams.

Sounds of marching, sounds of tanks moving, frightening sounds. Sounds of rockets being fired.

I really believed my life was in danger and asked myself if I was prepared to die for what I believed in. I had followed my coincidences, followed the synchronistic events which had brought me here. I had my signs along the way, the David Bailey sign at the airport, the slight breeze at the shrine and the cross of twigs on the lake and I had to answer yes, I was prepared to die for what I believed in, to face death for my God. I lay there frozen with fear in my bed, hands over the sheets in case the terrorists burst in, then there was a knock at my door. I ignored it. Frozen, frightened and stiff I laid there the rest of the night.

It was shortly after dawn that I heard a child running down the passage; it was Ashley's daughter and with that I knew that it must be safe and with that also I must be safe. A sense of relief washed over me and then I rested for a few hours.

I didn't know if I had slept or not but there was a further knock on my door and I was told that my breakfast was ready. Everything on the houseboat and outside seemed quite normal and calm. Again I decided to keep my experiences to myself and also decided not to ask any questions. Today I was going to Bernard's house for lunch.

Sunday

Bernard was a friend of Ashley's who sold locally made jewellery and religious artefacts from his house a few miles out of town. We journeyed there together in a small three-wheeled pop-pop taxi. Not knowing Bernard too well I didn't discuss any of last night's or yesterday's events.

Driving out of the town we passed some of the sights I had seen with Abdul two days ago. As we journeyed we drove towards the area where the government and local officials lived; this had been the area that was under attack by the terrorists yesterday. The main road they all lived on is protected by army guards with security barriers at either end and overlooks the lake. It is also a short cut to where Bernard's house is.

There was a small problem with me getting through the security barriers, maybe because I was a foreigner and I told Bernard I didn't want to go if there was a problem. Also, I was still a little nervous after the events of the past 24 hours but Bernard quickly had it sorted and we were on our way.

We stopped en route at the local timber mill. Bernard was having a new house built and he had some issues about wood. Soon after stopping Bernard and the mill owner got into a heated discussion. It appeared the mill owner didn't have too much respect for him. I left them to get on with it and shared a cigarette with an old man who was driving his horse and cart to the yard.

The old man's demeanour was typical of what I found with Kashmiri people, happy, smiling and open with great eyes. When Bernard had finished his argument we continued to his house.

Here he showed me many things that I could buy. He and his family also provided lunch. I wanted to buy a wall tapestry and asked the price. He said forty pounds and, because that was exactly the amount of money in sterling I had in my pocket, I agreed to buy it at that price, another synchronistic event I thought. His house was fine, quite modest but his new house was absolutely beautiful. It had three floors set overlooking the lake and the work inside was of a very high standard.

One thing that I noticed and noted was each door had really big locks on them, every door in the house. At the top of the house he had his own private prayer room also with big locks. In my jealous western way I wondered where he had got his new-found wealth from.

After my visit he arranged for his son to drive me back. His son was a nightmare driver who drove too fast and never stopped honking his horn. Like his father he also stopped along the journey and had a heated discussion with someone before speeding off at top speed, until we came to a traffic jam. He had intended to take the lake road back to Srinagar town centre but there was a problem so we turned around and he decided to take the road that the government officials lived on. As we did so I noticed a large billboard saying 'Love transcends all barriers'.

I asked who would have put the notice up and was told that it was the Indian Army.

We turned the corner into the officials' road but this time there were no security barriers and no army guards. The barriers had been rammed and buckled, the guards now gone. It seemed that there was a gun battle going on right now, right here.

As we drove quickly down the road I noted an Indian soldier, with bullet proof vest on and gun in hand, heading off to join the fight. This guy didn't seem too keen, with head bowed low and tiredness in his step. We sped down the road and headed off to the town centre.

In the town there were people running about, the sound of windows being smashed and to me a bit of an alarming situation. However, the Indian soldiers in the town didn't seemed too concerned, maybe they saw it as people letting off steam. We parked by the quayside and as I boarded the shikara I saw a few young lads in the front of the mosque cheering at the fighting taking place on the nearby hillside. Most people seemed unmoved and calm about the situation as did my wallah who gently and slowly paddled me back to the houseboat.

The plan for the next day was to go fishing with Ashley, so I decided to buy some beer from the floating shop that travelled about the lake. Relax on the deck, have a drink, have some dinner and an early night. In doing so I got into conversation with one of Ashley's friends.

We didn't directly discuss the fighting but he did raise two points that I found interesting. Firstly, he posed the question of who did I think was behind Osama Bin Laden? To which I said I wouldn't like to speculate. Secondly, he questioned whether you should take money from these people? Which raised the question for me, who are these people and has money been offered to some local people to support terrorism?

Feeling uncomfortable about getting into too much detail on a subject I knew nothing or little about, I changed the subject. He then asked me if I believed in God. I explained that for me there was no doubt that a greater power existed and interacted with the world, whichever name we choose to give that power, and that these views were not about faith or belief but were much stronger than that for me. After he had gone I went to bed hoping for a better night's rest.

Sleep proved difficult again, the ceiling fan's rhythmic noise was disturbing but at least I was managing to rest my mind.

Until, "We're so proud of you, so very, very proud of you... You farting, wanking, fat bloke, you farting, wanking, fat bloke."

Repeated over and over, almost sang in an insulting tone.

The voices were the same as I heard a few nights earlier but this time they seemed to be coming from somewhere a lot nearer than before. They seemed to be coming from the houseboat moored next door. After they had stopped I lay there, listening for what would come next. The more I listened the more I could hear the faint sound of something but it was unclear what it was, maybe just a TV set or radio and all the time the beat of the ceiling fan continued to unsettle me.

Then followed noises and voices that seemed quite clear and quite nearby, someone asked, "How we getting on with him next door?"

I could see people on the next houseboat and was certain that this was where the voices were coming from.

I tried to rest and to sleep but the voices and noises continued. At one stage during the night I got up, went on to the deck and shouted at the houseboat next door.

"God is important, Allah is important, you shouldn't mess with people's religions."

This fightback seemed to have an effect as the rest of the night was quiet and I got some sleep.

Monday

During the night I had decided not to discuss my experiences because I was in a strange country, with people I had only known a few days and wasn't sure who I could confide in. Although I trusted Ashley and Abdul, they were good men, I didn't feel I could discuss the voices with them. Also it was unclear if they were involved in any way.

In any event the following morning was bright and I took breakfast on the deck and watched the mothers take their children to school in their own small shikaras. There was one small boy about ten minutes behind everyone else, paddling as fast as he could and I imagined him getting in to school late and telling his teacher that he had dropped his paddle in the lake on his way to school, as a way of explaining his lateness.

There was the morning procession of shikaras selling things, the floating shop, the floating milkman and also, Mr. Wonderful Flower Man, from whom I bought some flowers for the ladies who had being cooking my food, which had been excellent.

Everyone seemed so relaxed, happy, bright and calm; it was a good Monday morning.

Ashley and his friends arrived and we headed off fishing. We were going to Pahalgam about 100 kilometres east of Srinagar to fish on the banks where the East and West Lidder Rivers meet.

As we drove east out of town we passed a convoy of about 50 Indian Army trucks. Ashley told me it was a daily convoy to Leh. The airport there was very steep to land into, so the troops stationed there were supplied by road.

Our first stop was at Avantiswarmi Temple, ruins of a Hindu temple over a thousand years old. We drove on, past saffron and paddy fields, past an area where they made handmade cricket bats, which were hanging from the trees. Through a small town with a ten Rupee toll, where they had been collecting money for years to repair the road but obviously not charging enough.

All along the road, every few hundred yards, there were Indian soldiers, some with mine sweepers checking the roadsides, some in fields, some just standing guard. It was a heavy military presence. This was because of the Hindu Pilgrimage of Yatra to Shiva's cave at Armanath.

We stopped along the journey to buy some bread and as we did so an army patrol was walking past. I reached across to open my window and the small movement of my hand caused one of the soldiers to cock his rifle. I thought these guys are on a high state of alert or maybe it was because I was a foreigner.

We arrived at Pahalgam were it appeared our guide needed a guide, which made me smile. Then we set up a base under one of the walnut trees. There were cow pats everywhere but at least it was Holy shit.

Whilst Ashley and his friend set up the fishing kit, I just took in the surroundings and chatted to a soldier patrolling the area. I listened to the flow of the two rivers as they raced down from the Himalayas and started reading a book, on the recent history of Kashmir, that I had bought on my day out with Abdul.

Time passed quickly with Ashley waist deep in water battling the currents of the river in an attempt to catch our dinner, me reading and watching as, first cows, then people waded across the rivers. Ashley's friends sat joking and relaxing, it was great. Soon it was time for lunch. Ashley's wife had prepared a curry and we all sat down together to enjoy it sitting on a blanket under a walnut tree.

As we did so Ashley started to tear off pieces of his naan bread and throw them away. I asked what he was doing and he said he liked to share his food and that the food he had thrown would do some good to some other living thing. I thought this a wonderful gesture, a habit I would like to copy. A shepherd with a herd of cows was walking past and Ashley invited him to join us, which he did. Two stray dogs came by; they also got some of the lunch.

After lunch we continued as before, all of us totally immersed and enjoying what we were doing, me reading and listening to the sounds of the rushing rivers. It was as I was doing this that I detected another sound just below the sound of the racing rivers. As I listened harder I could hear a hymn, I could clearly hear the words being sung and being sung well: 'God rest you merry gentlemen may nothing you dismay, for Jesus Christ our saviour is born upon this day, to save us all from Satan's sins and help us on our way, oh tiding of joy, comfort and joy comfort and joy, oh tiding of joy, comfort and joy.'

The hymn was repeated time after time. As I listened I considered were the sound might be coming from. Were there loudspeakers in the nearby trees, or was it God, or was it imagined, or was it something else? I was enjoying it whatever the source.

After a while the hymn changed to a different one. This time it went, 'Bread of Heaven, bread of Heaven, I will always give to thee, I will always give to thee. Bread of Heaven, bread of Heaven I will always give to thee... I will always give to thee.'

I sat, listened, enjoyed and wondered what was happening. These sounds were making me happy, unlike the sounds from the night before. Maybe that was because I was outside, relaxed and with people. Ashley was in the river fishing so I asked one of his friends if he could hear the hymn just below the sound of the rushing rivers. He stopped to listen and said maybe he could hear a faint sound but what he could hear if anything was a Kashmiri song.

I returned to my book and listened to the hymns, happy and smiling. Whatever it was, it was relaxing me and I was enjoying it. One of Ashley's friends brought me a freshly caught, freshly cooked trout, it was wonderful.

Having caught enough fish for dinner, Ashley checked with the fishing official, paid for the fish caught and we started our journey back. As we did so a pop-pop driver was driving his car into a shallow part of the river to wash it. On the way back we went via Mattan, to visit the Hindu temple there and feed the Holy fish. It had been a perfect day for everyone.

I went to bed that evening happy because it had been

such a perfect day. I laid my head on the pillow and hoped for sleep but the repetitive noise of the ceiling fan continued to hinder that; it seemed as if it was playing a tune, as its constant rhythmic hum played into the night. I again started to hear other noises, this time water noises, like what I imagined Chinese water torture to sound like. Then sounds of what sounded like a chainsaw. Sleep was going to be difficult again.

After a while I turned on the light to search for the control of the ceiling fan and turned it off. This helped for a while but the noises and voices came back again. I got up again and looked across at the houseboat moored next door. I felt sure that was the source.

There was a young man using a pole to clear away the weed from the side of the houseboat but it was the middle of the night. I sat at my window pouring water from my water bottle into the lake, in an attempt to show I was not frightened, but of course I was, increasingly so.

Then going back to bed and trying to sleep I heard, "Ooooooooooo-d-I-o, bang, bang, bang."

The sounds were sung like a song but constantly sung over and over, and only just audible.

"Ooooooooooooooo-d-I-oooo, pow, pow, phow, ooooooo oooooooo-d-I-oooooo bang, bang, bang, ooooooooooo-d-I-oooooooo, you're a wanker, you're a loser ooooo-d-I."

And so on. Right thought the night, as if it was on a constant loop, with different words and different emphases changing all the time, interspersed with water sounds and chainsaw sounds, then followed the dawn; now the sound of the dawn and the sound of the crows landing on the tin roof started to scare me.

Tension was all around my neck and I was starting to find

it difficult to think straight. Having got no sleep, I stayed in bed, missing breakfast, not coming out of my room till mid-morning.

Tuesday

When I got up I found Abdul waiting for a shikara to take him into town to go shopping. I asked if I could go with him because I had decided to buy a small transistor radio with an ear piece in an attempt to shut out the voices and sounds. I felt safe with Abdul although now I was starting to feel tired from lack of sleep.

We set off together. First stop was the market, then on to a bakers where they baked wonderful Kashmiri bread on the sides of an oven in the ground, after which we found two small electrical shops, the second of which had just the type of radio I had in mind. Abdul ensured I didn't pay too much for it. Finally we stopped in a small café in the market for tea and chips. I liked being with Abdul and teased him that he was like my dad. He got a little upset seeing us more like brothers but we were close as human beings and someone who I will never forget.

It was early afternoon by the time we returned to the houseboat and I decided to try and get some sleep or at least some rest. There was a boy on the houseboat next door fiddling with a transistor radio which started to grate on my nerves. Now I started to get annoyed and maybe a little paranoid but didn't feel able to say anything, just lying on the bed and resting until dinner time. While resting I had decided to tell Ashley about the voices and the sounds and the whole experience because I was starting to need to talk to somebody about it, feeling increasingly uncomfortable about it and increasingly frightened about it.

Ashley unfortunately was not around that evening having gone out somewhere. In need of an outlet I turned on the small TV but the only station I could get had a very poor picture and very poor sound quality which only annoyed me and therefore added to the stress that I was feeling.

With no one to talk to, with tiredness and with stress, I went to my room to try and sleep, but to no avail. The voices started almost straight away. Taking my new transistor radio and putting in the ear piece I hoped for respite. It didn't happen; now with my ear piece in and hiding under the covers I was still unable to find rest. The only radio station that I could find was singing the praises of Hizbella and the only other thing I could find was white noise, lots and lots of white noise.

Then the voices started again, "Ooooooooooo-d-I oooooooooooo bang bang bang. Ooooooooo-d-I-Oooooooo phow phow phow."

Again they continued right through the night. I thought about leaving the next day as I couldn't take much more of this. That would mean I didn't get to meet the Sufi Baba, which would be a big disappointment.

As I was thinking about the best thing to do, the voices started again, this time with a different chant, "Wouldn't it be a good idea to leave tomorrow, leave tomorrow, leave tomorrow?"

Sung, over and over, in a mocking tone.

"Wouldn't it be a good idea to leave tomorrow?"

Unsure whether it had been my idea first or because of their encouragement, in any event my mind was made up;

first thing in the morning I would see Ashley and tell him I needed to leave. I tried unsuccessfully to sleep, just lying there for the rest of the night waiting for the dawn with the constant sounds and constant voices, which only added to my already high tension levels, my already high state of fear.

Wednesday

The morning came but so did the noises and voices with a continuous low hum of, "Oooooooooo-d-I-ooooo."

Finding Ashley and taking him to one side, feeling certain now that the sounds were coming from the houseboat next door and not wanting anyone on that houseboat to hear, I told him about the voices, told him about my experiences. I told him that I knew myself well and, although not yet paranoid, I needed to leave. He listened intently, apologized for not being there the previous night and said he would try and book me a flight to Delhi that day, if any were available. Abdul came over and asked what was happening, Ashley retold the story to him.

Abdul said he was very disappointed I was leaving but he couldn't understand why I would say that the mosque was saying things about me. I explained that I wasn't saying the sounds were coming from the mosque, only that I could hear sounds.

Waiting for Ashley's return, I was growing increasingly tense, increasingly frightened, feeling increasingly vulnerable, the voices and sounds going on and on relentlessly.

Ashley returned after what seemed like an age and said

he had managed to book my flight to Delhi and we would leave in an hour.

I hurriedly packed my rucksack, as the sounds and voices got slightly louder and louder. The pains in the back of my neck got sharper and sharper, my heartbeat growing faster and faster.

Then, "We never lose a rabbit, we never lose a rabbit."

Repeated time after time and time again.

"We never lose a rabbit, you never lose a rabbit."

Now so many different thoughts came into and out of my mind. Tension levels up still higher. This was only increased further as the voices chanted, "Ouch and one will come too, one will come too."

Repeated over and over.

"One will come too."

Abdul, seeing my stress, came over and tried to comfort me saying that maybe he too could hear something. But any relief I found in his kindness was shattered by, "OOO-Di-Ii-O, you're a wanker, you're a creep, we're going to kill you, we're going to cut you, we're going to rape you, you're a dead man, you're a dead a man, we're going to hurt you, we're going to kill you, you're a dead man."

Sung, almost chanted, again over and over.

Then there was a woman's voice, she spoke clearly and she said with a stern tone, "You're a dead man walking Andy Cole. You're a dead man walking."

Being unable to take any more and fearing I might not even get to the airport, let alone Delhi, I found Ashley and told him that I needed to ring the British High Commission, wanting to let someone know of my situation and condition.

We left for the airport shortly after and on the journey I could tell Ashley was picking up on my fears and anxiety and, with that, started to feel tense himself. However, he ensured I safely arrived at the airport and that was where we said goodbye.

As I checked my luggage through security and entered the terminal I thought I could still hear the faint sound of, "Oooooooooo-d-I-ooooo" in the background.

It was after I had found a seat inside the terminal and started relaxing a little and calm down a few degrees and listened to the piped music that I heard the sound of the woman's voice again, this time she clearly said, "When you go take your life with you, go and take your life with you."

Confused, I struggled to understand what was happening. It seemed that the voices now had access to the internal loudspeaker system, inside the airport. This only added to my concern, added to my confusion as I battled to come to terms with the situation. But at least they had told me I was going to live. Then a male voice clearly said in a taunting tone, "Enjoy your last night, enjoy your last night, wherever you are we will find you."

Mentally exhausted, I boarded the plane and waited for take-off, listened to the piped music, and felt a little safer than I had done in a while and believed that my ordeal was now over.

During the flight I glanced over at the woman in the next seat, across the aisle from me and asked if I could borrow her newspaper. The headlines said that there had been a car bomb, in Srinagar, on Monday and although under maximum security a high ranking government official had been killed. It seemed that the terrorists had got their target of the previous weekend.

It was just as I had finished reading the article that the voices started again, this time a man's voice clearly spoken said, "We normally blow these things up."

Looking around it appeared that nobody else had heard what was said, if so no one reacted to it. My mind started racing again.

"Yes, that's right; we normally blow these things up."

What was happening, where were these voices coming from? Was the plane really in danger of being blown-up?

Shortly afterwards the woman's voice said, "You can never go back to Kashmir."

This was soon followed by, "Why don't we just pull together, ugh pull together, ugh pull together, ugh?"

This was sung by both a male and female voice.

Then the woman, on her own again, this time spoken, clearly spoken, said, "If we did pull together, you may be able to return, this was the only way you could go back."

Nobody else appeared to be hearing any of this.

The woman across the aisle from me dropped what appeared to be a piece of rubbish. The male voice said, in a commanding tone, "Pick it up, pick it up."

It was an instruction to me. All I could do was follow his instructions. It was a ticket of some type.

"Put it in your pocket, put it in your pocket," was the next instruction.

Then the male voice, again on his own but in an excited manner, said, "Boom, boom, boom, let's blow up the moon. Boom, Boom, Boom, let's blow up the moon."

By this time my mind was all over the place but we would shortly be landing.

After landing and taxiing to a standstill, what seemed like an announcement came over the airplane's intercom system, "Could all those passengers holding tickets from Srinagar please stay on the plane until its next destination."

Where would that be? I thought to myself.

"Just stay on the plane," the female voice said.

I decided to get off while I could and boarded the courtesy bus to the terminal. When the doors were closing on the bus, the female voices said, "Bing bong. We must learn to do what we're told, mustn't we?"

What was happening, how could the voices be everywhere?

Entering the terminal the voices gave a further instruction.

"Pick up the package and get back on the plane, just pick up the package and get back on the plane, use the ticket in your pocket pick up the package and get back on the plane."

Scrambling in my pocket I found the ticket that had been dropped by the lady in the seat across the aisle from me on the flight. Taking the ticket I ran to various desks to ask if they had a package for me, showing them the ticket. Every one of them looked at me blankly.

The more I failed in my task the angrier and louder the voices got.

"Just pick up the package and get back on the plane, it's not hard is it, just pick up the package... just pick up the fucking package, PICK UP THE FUCKING PACKAGE... too late OK... forget the package just get back on the plane, just get back on the plane... get back on the FUCKING PLANE... This guy's taking the piss. THIS GUY'S REALLY TAKING THE PISS. Get his daughter in... Cut off her left nipple..."

"SCREAM."

Now I was right on the edge of sanity.

I went to an army guard by the exit and asked for help but he didn't understand me, then a man in a white coat came running over and said, "Loco, Loco?"

"Yes," I replied.

"Wait here," he said.

Not trusting him and imagining being locked up in some Indian nuthouse, I left the airport and wandered down the road with no idea where I was heading. I was stopped by a taxi driver, who asked if I wanted a taxi.

"Yes," was my reply, "British High Commission."

I got into the taxi and the voices started again with, "Ooooooooooo-d-iooooooo."

If indeed they had ever stopped.

Somehow the sounds continued as we drove. It seemed like all along the roads of Delhi there were loudspeakers from which the sounds were coming. Totally confused, beyond guessing what was happening. Now the sounds were becoming like a backing track to what was being said.

"We've killed all your family... well not all your family, just Sheila, she's dead, our men in Australia got her, not Sharon, Sharon wasn't in, we'll do her later, but your dad he's dead, Michelle she's dead, Lin too she's dead... Can't find Helen, she's in Canada but our people there will sort that out soon... But the rest are dead... Now we're coming to kill you."

At that exact moment, at that very moment, the traffic lights failed and the taxi came to a halt and we became stuck in a traffic jam.

Taking all the money out of my pocket I gave it to the taxi driver as a fare, got out of the taxi and with my arms held in the air said, "Go on do it, do it now."

Not out of bravery but because of not being able to take any more.

I walked down that road in Delhi, for a few hundred yards, hands in the air, believing that this was it, really believing that this was it. The traffic lights must have come back on and the taxi I had been in pulled alongside and the taxi driver said, "Get in sir, please get in."

God bless that taxi driver.

By the time I arrived at the British High Commission my mental state was far from clear. The background sounds and voices seemed to continue as I told Mr. Kumar and his colleague that my family had been killed by terrorists. The first thing they did was to ring my family in England to check that they were alright, which they were, thank God.

Then they questioned me about what evidence I had regarding the terrorists, the voices and the package. It was as I placed my airline ticket to Delhi and the ticket that was dropped by the lady on the plane on the table as evidence, that I knew my story looked crazy.

Mr. Kumar asked if I wanted to go to hospital; my answer was yes. He said he knew one that would be safe and called the East/West Rescue Centre. Then they asked about my luggage, which had been left at Delhi Airport, and arranged to have it picked up to avoid a security alert.

With that he organized a High Commission car to take me to the East/West Rescue Centre. On arrival the doctors carried out some basic medical tests before leaving me to rest for a while. As I lay there resting the voices continued; again it seemed as if they were coming from the building next door.

Totally confused, totally mentally and physically exhausted, with a male nurse inserting a drip into my hand and with the voices chanting, "We're going to rape you, this is the man who is going to do it and you can do nothing about it, we're going to rape you, we're going to rape you."

Beyond caring, beyond thinking, beyond fighting, I slept for five days, waking only occasionally to identify my luggage, for more tests, for food, the toilet and to take a phone call from my dad, during which the voices taunted me with, "Andy Cole loves his daddy, Andy Cole lost his dad."

After I had been sleeping for five days, Mr. Kumar came to visit. Now, as I felt well enough, he arranged for my travel to the airport and my flight back to London.

As I checked out of the East/West Rescue Centre another English man checked in.

He was from King's Cross, London.

England

Back in the UK

Arriving back in England, still shell-shocked by my experiences and having slept most of the flight I was met by Michelle and my dad, who were both very concerned about the situation following the phone call from the British High Commission.

They took me back to my dad's flat in Thame, where I slept for a further 18 hours.

On waking the first thing they wanted me to do was to visit a local doctor to get myself checked out.

The doctor listened with interest while I told her my story, at the end of which she gave me her diagnosis. She told me that just as it was possible to have visual hallucinations it was also possible to have audio hallucinations, saying that these can be brought about by culture shock amongst other things and suggested sleep, rest and relaxation.

That's exactly what I did, spending the next two weeks with my dad.

We would spend time watching the cricket and the racing together. England on their way to beating Australia and winning The Ashes for the first time in many years and a horse trained by Martin Pipe called 'Getoutwhileyoucan', winning as it liked at some West Country racetrack. In his love and care I made a speedy recovery.

Some evenings from his balcony I would watch the sunset over St. Mary's Church with the church bells ringing out over Thame town centre, enjoying the beauty and thinking about the similarities with the sun setting over The Throne of

Solomon and the call to prayer echoing across the lake in Kashmir.

At night I would read *The Prophet* by Kahlil Gibran from a volume of his greatest works; it was a book I had bought on my first day out with Abdul whilst there. Sometimes at night I thought I could hear the voices again and felt a little vulnerable but this soon would pass. I had been happy to accept the doctor's diagnosis of audio hallucinations, although not totally believing it, as I struggled to understand my experiences.

One day I went on the bus to nearby Aylesbury. I had wanted to send a letter to Ashley to thank him for both looking after me whilst in Kashmir and helping to ensure I arrived home safely. During my time with him I had developed the deepest fondness and respect for him.

Whilst typing the letter in the library two young girls, both with hijabs on, came and sat down next to me. I greeted them with "Salaam Alaikum."

They looked back at me, rather surprised but replied, "Alaikum Salaam."

It means 'May peace be with you – and may peace be with you also'. I had grown deeply fond of Kashmir whilst there; it had touched something deep within me.

It was during this time of recovery that I noticed that the Amnesty International Lecture was being held at the Sheldonian Theatre, Oxford. I had graduated there about ten years previously and decided to attend. The lecture was entitled, 'War/Terror', and was too focused on the either/or line between them and how the line between war and terror was getting thinner and thinner, in the view of some academics.

After about two weeks and feeling fully recovered, I

needed to decide what to do next; should I go back to Gibraltar or continue my travels and go on to Australia?

It was in this uncertain frame of mind that I said goodbye to my dad and returned to Bushey, while seeing in which direction I would be taken. There I would be able to spend time with Michelle, my eldest daughter, and my ex-wife Lin. Money was an issue but only in the sense that maybe I could not afford to go to Australia.

During our time together Lin and I decided to visit Watford and while there we came across a DVD entitled *24*. This was apparently a TV series that she really enjoyed. Having lived in Spain for the past seven years, I had never heard of it. Lin enthused about it with such a passion that I had decided in my own mind to buy it for her as a parting gift.

We returned home and decided to take Lin's wheelchair-bound mother Pat out for lunch at the local golf club. As we left the house a car pulled up and parked on the pavement, thus blocking our path. We all looked at each other and complained about how thoughtless the driver had been. At the precise moment we were considering a different route, Michelle returned home with a present for her mum; it was the DVD of *24* that we had seen and discussed earlier. So I took time to explain that was a simple example of exactly the type of synchronistic event that I had been noting, experiencing and following.

Then, just as we finished talking, the car that had blocked our path drove off, thus clearing our way. With that we walked to our destination.

It was a Wednesday and because of finances I had decided to return to Gibraltar on the following Saturday. That was until my sisters, Sharon and Sheila from Australia, rang to say that they had a gift for me and that the gift was a return flight to Australia. What a wonderful surprise! Their gift had ensured that my travels could continue and with that I made my plans for the trip.

This also enabled me to spend an extra week with Michelle and Lin, during which time I arranged my visa, an ETA which allowed three months stay in Australia.

In addition I enjoyed some nights out, including going to see *Charlie and the Chocolate Factory*, and finish reading *Dawkins' God* by Alister McGrath.

Off To Australia

My sisters, Sheila and Sharon, had arranged a direct flight from London to Brisbane with a few hours' stopover in Hong Kong. The plan was for Sharon, my younger sister, to meet me at the airport, where I would spend some time with her and her family in Brisbane.

Then for me to go off travelling around Australia on my own, doing some research into Aboriginal spirituality because to me it seemed clear that if God, Allah, a greater force existed, then there would be evidence of that existence from before the time of the Prophets and the indigenous Aboriginals' ancient culture could possibly provide pointers towards that existence.

The journey was to end in Sydney where I would spend time with Sheila, my older sister, and her family.

The highlight of the flight was the stopover in Hong Kong, sharing and swopping cigarettes with some Chinese men in the smoke-filled smoking rooms, generally exploring the airport and chatting to fellow travellers.

The second leg of the flight passed quickly and Sharon met me at Brisbane Airport with her husband Aussie and son Zane; her daughter Rhiannon would join us at their house, later. We drove through the town, my first experience of Australia, passed the Ekka Show Ground, passed Brisbane Cricket Ground, seeing the Yellow Cabs company taxis, which are really orange and taking in some of the other sights en route.

Their house was about 30 kilometres outside the town centre in a suburb called Beenleigh and when we arrived Sharon gave me one thousand dollars, which was very timely and would help to fund some of my travels. We sat and drank some wine and chatted for a while before I crashed with jet lag and slept for 18 hours.

The next day was a Sunday and it had been decided that Sharon, Aussie, Zane and Rhiannon would take me on a local tour. The journey would be my first taste of the Australian countryside, which was wonderful. First stop was a pub for some beer and to watch the cane toad racing. Then we headed off over Tamborine Mountain with its winery, spiritual healing centres, rain forest and fantastic views, towards the coast.

We were heading for the Gold Coast; our particular destination was Broadbeach, which lies just south of Surfers' Paradise. The beaches are very clean and open with the current in the sea very strong, as the waves travel many thousands of miles before crashing into the shore. A small market was taking place and ibises, a common local bird but one new to me, were all around.

For lunch we went to an RNLI club house on the beach. It had great views over the ocean and was a gambler's dream with multi-sport TV screens, Tab and Pokies (gambling machines) everywhere, all supported by a bar and restaurant. It struck me how big gambling is in Australia and not only catered for but actively encouraged. The lunch, of mainly fish, was excellent but a personal situation developed between Sharon and Aussie which meant that he would be going to stay with friends that night.

That evening and over a bottle of wine my sister and I sat and spoke; it was the first time we had been alone together for over 30 years, indeed I had only seen her once in the last 16 years. However, the bond between us was still strong and we were close. It was during the conversation that she told me that her marriage was over and she and Aussie would be separating soon, which to my mind was a great shame as

Aussie was a very good man, a top bloke. We also discussed different ideas on spirituality and religion and exchanged different book titles for each other to read. I told her a little of my experiences in Kashmir and India without going into too much detail; we both had a good evening together.

The following day Aussie returned and he, Zane, Rhiannon, Sharon and I relaxed in each other's company as they all helped to plan my next day's trip, which would be to Brisbane, to see the city.

My first impression of Brisbane was just another city but the more I explored it, the more it started to grow on me. Opening my guide book I decided to take one of the recommended walking tours, starting at South Bank Parklands, visiting the Queensland Art Gallery, Queensland Museum and CBD before finding the City Botanic Gardens and its mangroves. The more I explored, the more I liked the city. Some interesting things also happened, for example.

Whilst sat by the river an old paddle steamer called the *Kookaburra Queen* came up the river and I wanted a better view of it. The boat came steaming up the river to just past were I was sitting, then started turning, giving a full and thorough view of itself. I considered this an interesting coincidence, as what I had wanted had just happened.

Also, whilst visiting Borders book shop, there was a particular book that Sharon had recommended that I wanted to buy. After searching high and low I was about to give up. As I turned to leave I noticed the book that I wanted, right in front of my eyes.

From here it was back to Queensland Museum to see what I could find out about Aboriginal culture. Unfortunately the museum didn't have much on display, just a few Aboriginal artefacts. However, the receptionist was very helpful and recommended works by Mountford, Spencer and

Gillen, Berndt and in particular a book by A. P. Elkin called *Aboriginal Men of High Degree.*

My research into the indigenous Aboriginal Culture had started.

I was to spend the next four weeks with Sharon and her family, getting to know them better and enjoying their company. Zane was 12-years old, very sporty and a deep thinker; Rhiannon was 14 and a real gem, full of life and fun. Their friends would often come to visit and we would spend time feeding Bella, a tame wallaby who would come for her tea and would feed out of the hand, even with her joey (baby) in her pouch.

We would play Monopoly together and, oddly, the first game we played, all four of us threw the same numbers, twice, as we vied to see who would go first. The odds of this happening are many thousands to one.

Aussie introduced me to the Pokies machines, which I would become very familiar with during my trip, in particular, 15 Lions, and also to Google Earth, which used interesting satellite technology. For my part I would spend time reading, thinking and enjoying being with them all.

At the end of the first week we went to watch Rhiannon playing basketball; the sports centre was packed with parents watching their children playing sports, very healthy and very Australian, after which Aussie took me to meet his dad, Zane (Snr.), who lived in an amazing wooden house that he had been renovating. He showed us a little nest in the roof that he had built for a wild possum who was sleeping inside. Some butcher birds came by to ask for food; they just flew into the kitchen and started singing. Zane (Snr.) took some cheese, threw it into the air and the birds jumped to catch it. Parrots flew about in the garden and I thought how friendly the wildlife was in Australia.

During the second week Sharon and Aussie were working, Zane and Rhiannon were at school and I read, studied, thought and explored.

For the second weekend a trip and a few days away had been planned to Currumbin, to the south of the Gold Coast. An apartment had been booked on the beach with great views overlooking the sea. We were to enjoy a few day here, with friends of Sharon and Aussie coming to visit, play and rest. Again the wildlife was friendly with a magpie coming into the flat and a local cockatoo that came by every day at four o'clock to be fed.

On the beach was a very large rock and you could look back towards the centre of the Gold Coast with the rock in the foreground and the city in the background, which gave a beautiful stark contrast between the beauty of nature and the beauty of modern man.

At night the sea was relentless and powerfully pounded the coast continually, as if it were a never-ending generator. During our time away it was Sharon's birthday and we all went out for dinner at the nearby RNLI club. Aussie and I played the Pokies, 15 Lions, and won the jackpot, which brought our winning for the week to well over a week's wages each.

The third week was spent going through my mum's belongings. She had died in Australia three years earlier and Sharon had been unable to face going through her things alone. As we did so we came across letters, pictures of our childhood and various other things of personal interest that our mum had saved over the years and Sharon recalled how Zane and Rhiannon had called her 'Har', being unable to say Grandma.

It was also during this week that Aussie had arranged a visit to the nearby Jupiter's Casino, where, coincidentally, they were holding their first ever night of live Texas Hold-em Poker, a game I had played most weeks while in Gibraltar. It was a good night and again we increased our winnings on the 15 Lions Pokies.

The final week in Brisbane was spent planning the next leg of the trip, which would be to Darwin with a view of maybe trying to get to Arnhem Land, an Aboriginal settlement before heading to Ayers Rock and beyond.

It was also the time of the Brisbane River Festival and tickets had been booked for Zane and Rhiannon to go to see *Charlie and the Chocolate Factory*, the last film I had seen in England with Michelle and Lin, so while Sharon took the children to see the film with some friends, I explored some more of the city.

The festival itself is a yearly event, the highlight of which is a large firework display over the river, set to music, which is both opened and closed by a fly pass of a RAAF F1-11; the sound of the plane as it flew low overhead is frightening and as it did so the pilot turned on the afterburners to the delight of the crowd.

The final evening in Brisbane was spent having dinner at a friend's house where we ate like kings, drank like fish, got high on some home-grown marijuana and laughed till it hurt: the hospitality of Sharon and Aussie's friend was fantastic.

It seemed clear to me, reflecting on the time with my sister, Sharon, and her family, that the synchronistic events and coincidences were still happening and therefore I was still on my right path.

Off To Darwin

The flight time from Brisbane to Darwin is about four and a half hours and is a testimony to the sheer size of Australia; in that time you can fly completely across Europe but only about half way across Australia.

During the flight I got chatting to my fellow passengers. One was a nurse en route to take up her new duties at Darwin Hospital and begin a new life, the other a local man on his way home to a book launch; he told me some detail about Darwin. For example that the locals call it the Top End, that it only has two seasons, not four as in Europe, wet and dry, and that it was unsafe to swim in the sea because of the jellyfish and the salt water crocodiles.

The first thing that hits you as you land is the heat, as it's about ten degrees hotter than in Brisbane but the airport was well organised and from a selection of hotels and hostels I chose the Value Inn Motel as my base.

The first evening was spent exploring and visiting some tour operators, having dinner at a local pub called Shenanigans and ending up at Darwin Sailing Club, which was an absolute bonus as the sunset over the sea from here was magnificent. The enormous sun slowly sunk below the sea-covered horizon, while I sat and drank a glass of Chardonnay, beautiful.

After the sun had set I took a taxi back to the town centre; the taxi driver was from near Kashmir, and I decided to check out a bar in the Backpackers, adjacent to my motel. This was to be my first experience of a Backpackers; it was called Melaleuca on Mitchell and I was very impressed. There was a large bar area with friendly staff, good music, a swimming pool and a kitchen to cook your food in. Getting into conversation I decided to stay for a few drinks.

I got into conversation with a rodeo rider called Ponjo who told me some of his stories of how he and his friend would go rustling. He told me stories of clear skin bulls, bulls that have never seen humans before and how he could tell they were clear skins by the look in their eyes.

He told me about the bull catcher vehicles used to track, chase and catch the wild bulls and how rustling was possible because the outback stations were so vast they couldn't be policed. Also he spoke of a friend of his who had died recently in such an endeavour. He explained in detail how they chased and caught the bulls and also told stories of his rodeo exploits.

This was my first conversation regarding the outback and I found the whole conversation fascinating. He also gave me his views on the indigenous Aboriginal and said whole pieces of land bigger than England had been given back to them. Eventually he decided to go off and find some women to talk to and as he did so I noticed a slogan on the back of his shirt that read 'Born to ride', an obvious double entendre, I thought. I decided to have one more brandy and then headed off to bed.

The next day was a lazy day sleeping late and having a wander around Darwin town centre, exploring one of the beaches. The beach had a notice explaining that this was Aboriginal land and as they were the hosts they had a duty to welcome visitors, while the visitors had a duty to respect the land; this I considered very honourable.

I also considered what to do next, should I go to Arnhem Land, Kakadu National Park or a trip to the Tiwi Islands, all of which would have been great experiences but as the budget was tight I decided instead to visit Mindil Beach Sunset Market.

The market was great with all the hustle and bustle you would expect; stalls selling food from many different Asian countries, music provided by some different local groups and a whole range of goods for sale. Oysters were cheap so I decided to treat myself to a half dozen Oyster Kilpatrick, sitting and enjoying them while watching the eagles playing on the currents of the wind on the edge of the beach. There was a Baha'i faith stall and as I had seen the Baha'i temple in Delhi I decided to take the details of the temple in Sydney and made a mental note to visit when there.

As the sun was now setting I decided to wander onto the beach to enjoy it; to my surprise about 300 other people had exactly the same idea, and the beach was packed. As the sun started to slowly set or, rather, as the earth started to slowly turn away from the sun because that's what's really happening, I was struck by the spiritual need within and considered that, possibly deep within us all is a deep spiritual longing, a deep spiritual questioning.

Time for one last look around the market and a wry smile at the stall that advertised 'Road kill, you kill it, we grill it', which was selling sausages and steaks made from camel, kangaroo, crocodile and other local animals. I thought I would give that a miss and then headed off to the MGM Grand Casino, which was a short walk away, quickly lost more than I should have done on the Pokies, 15 Lions and decided not to play them anymore without Aussie.

The following day was spent visiting the Museum and Art Gallery of the Northern Territory. Their star attraction is Sweetheart, a stuffed salt water crocodile. He was caught in 1979 and died soon after. Testament indeed to the caring nature of us civilised human beings and I considered if the Aboriginal people would have done such a thing. That is, take a beautiful creature out of its natural habitat and then, when it dies because of those actions, proudly stuff and display it.

The museum also housed details of Cyclone Tracy which destroyed half of Darwin on Christmas Day 1974. Interestingly, the Aboriginal Land Act came into force less than two years after Cyclone Tracy and I considered if the Australian Government had in some way acknowledged that a power stronger than humans pervades the planet and therefore sought to make recompense for past actions towards the indigenous population.

From here it was a walk along the seafront through the Botanic Gardens to Cullen Bay. Here I found a fantastic restaurant which overlooked Fannie Bay and served great prawns, oysters, mussels and steaks, all eat as much as you want and all enjoyed again with a majestic sunset; it had been a great day.

This was to be my last night in Darwin. I had decided to leave the next day and take the Greyhound bus south towards Sydney.

My ticket allowed me to stop whenever and wherever I wanted but I couldn't turn back, my first planned night stopover would be Alice Springs.

The lasting memory of Darwin for me is both the magical sunsets and the wonderful food.

Into The Outback

The journey from Darwin to Sydney is 6,000 kilometres. There was no time limit on my trip but I couldn't turn back. I had planned to stop along the way to explore and learn. As I boarded the Greyhound bus early in the morning I noted that U2 were playing on the bus radio, 'Sometimes you can't make it on your own'. After about 20 minutes the bus broke down and we all awaited a replacement, before we were able to set off proper along the Stuart Highway towards Alice Springs. An alternative to taking the bus would have been to use the Ghan, a railway with lots of history but the Greyhound bus offered greater flexibility.

Our first scheduled stop was Adelaide River, about a hundred kilometres from Darwin and, as we approached the town, the bush was all blacked from what had obviously been some bush fires in the area. In the event of a major fire the small local fire station, which was situated on the left as we drove into the town, would have found it difficult, if not impossible to cope. From here it was on to Pine Creek, famous for its bird life where a book-studying ornithologist got off the bus. Onwards along the Track, or Stuart Highway, towards Katherine, a town that had suffered severe flooding in 1998; pictures of the floods adorned the wall of the café we stopped at for lunch.

After lunch I went outside for a cigarette and decided to take some shade under a eucalyptus tree. As I approached the tree most of the birds flew away but one remained. As this bird seemed quite friendly I decided to start talking to it, to see if it would respond.

So I starting saying "You're pretty, you're pretty" to it in a repetitive way, "You're pretty, you're pretty". Before long the bird starting chirping back with what sounded like "You're pretty, you're pretty". We, the bird and I, communicated with each other for a few minutes. Then, just as I started to walk away, all the other birds returned to the tree and they all started chirping. So that as I walked to get back on the bus it sounded as if the whole tree, now full of birds, was chirping "You're pretty, you're pretty". The whole experience made me feel deeply happy and I said hello to my God.

A new passenger joined the bus at Katherine, a woman, about my age, attractive. We didn't speak but we were both aware of each other; she took the seat in front of me.

As we drove further south along the Stuart Highway we started to pass termite mounds, lots and lots of termite mounds. Mile after mile of termite mounds. At first I thought there must be thousands and then recalculated that there must be millions. The mounds were all various shapes and sizes and as we drove further south the termite mounds could be seen as different figures, some with faces, so much so that someone had stopped to paint faces on them as if to highlight the point.

As we drove continuously along you could more clearly see the figures, a Buddha, a wise man, a Holy man, a fat Michelin man, and all types of different figures. The continuous sound of the bus's wheels on the road, the motion of the bus and the figures of the termite mounds all combined to create almost a trance-like state. It was now, at this point that the voices started again with, "We know where you are, Har, we know where you are, Har, we know where you are, Har, we know where you are, Har."

Said, again almost chanted, but this time to the sound and rhythm of the bus's wheels on the road. Immediately I recognised the voices as those that I had heard while in Kashmir and India and started to sweat, started to get nervous, started to feel vulnerable. But the voices went on: "We know where you are, Har we know where you are Har."

Har was the name that Zane and Rhiannon used to call my mum, being unable to say grandma. How could anyone know that, but still the voices continued with, "We know where you are, Har, we know where you are Har."

Occasionally interspersed with, "We love you Andy Cole, we love you Andy Cole."

Then back to, "We know where you are, Har, we know where you are Har."

All in tune and in time with the sound of the bus's wheels on the road. Frightened and concerned but not feeling as vulnerable as when first hearing the voices in Kashmir and India I decided the best way to try and cope with the situation was to try to get involved with something, so that the voices didn't invade my hearing and thus didn't invade my reality.

It was about this time that we pulled in to Mataranka for a break, so I took the opportunity to get into conversation with the woman who had got on the bus at Katherine and taken the seat in front of me. She told me her name was Michelle, same as my daughter's and that she was from Holland.

Between Mataranka and Tennant Creek, there was one other stop at Daly Water and that part of the journey was spent getting to know Michelle and a man from Japan, who was sat close by, which meant by the time we arrived in Tennant Creek the three of us had formed a nice little friendship and I recalled the song that was playing as I first boarded the bus, 'Sometimes you can't make it on your own', and said thank you to my God.

Tennant Creek had the strangest licensing law that I had ever came across. We arrived a little after 9pm and decided to buy some wine to have with dinner but found that it was illegal to buy wine or any take-away alcohol after 9pm but you could buy sprits or beers if drank on the premises, the exception to this is on Thursday when it's no booze day, all day, and illegal to buy any wine.

It appears the local authorities had introduced the legislation because the local Aboriginal people, who only drink wine, draw their dole cheques on a Thursday and the law stopped them getting drunk and spending all their money in one day; the reason for the 9pm curfew is to keep the town quieter and safer at night. It seemed to me little wonder that some local Aboriginals drank because we, the white man, had massacred thousands of their people, taken their land and destroyed their culture. As it was the three of us managed to find a pub and had some beers and pizza for dinner before boarding the bus for the last leg of the journey to Alice Springs.

Alice Springs

We drove through the night, arriving in Alice Springs at about 5am, after the 23-hour bus journey from Darwin. Occasionally along the way I thought that maybe I could hear the voices singing, "The wheels on the bus go round and round," repeated to the beat of the noise of the bus but was unsure if the sounds were real or imagined.

The town centre was still busy when we arrived, with police and people everywhere; it had, as the bus driver remarked, "been a busy night in town". It was the weekend of the Henley-on-Todd Regatta, during which boats raced down the dry River Todd, with the crews' legs sticking out underneath the boats. This event is supported by the Harley Davison parade, where hundreds of Harley Davison owners parade their bikes around the town centre. A busy night indeed.

Sunday

The three of us decided to check into a Backpacker across the road from the bus station called Melanka; it would be my first experience of sleeping in a Backpackers. I waited while Michelle assembled a bike; it appeared that she was cycling around Australia, an incredible thing to do but had decided to take the bus for the journey between Katherine and Alice Springs. Her next destination was Coober Pedy, where she and her ex-lover had enjoyed some special times together but she had now lost her man to a younger model and was reliving her past dreams.

We checked in. The receptionist, who was also Dutch, and Michelle exchanged some words in their native language before Michelle and I decided to share a four-bed dorm together. We had breakfast and talked some more while waiting for our room to become vacant, and then set about exploring the town.

We wandered along Todd Street with its Sunday craft market to Aztec Hill from where we enjoyed the views of the surrounding mountains. As we walked and talked it became clear we got on very well together, talking about a whole range of subjects. Then we headed back thought the market, stopping to talk to a lady who had written a book about how her family had come to settle in the area and was now selling her book from her market stall.

Then on to a café for some tea and as we did so the local town crier walked past, wearing a traditional Afghanistan outfit so we asked why he was dressed that way. He said that he had wanted to express in his role as town crier those things that had helped to shape and build Alice Springs and as the Afghanistan camel drivers had played a leading role in its early life and development, wearing traditional Afghanistan dress for his official duties was his way of acknowledging this.

From here it was back to Melanka past a bar called Bojangles that has internet satellite viewing, which means that by logging on to their internet site your friends can see you in the pub via any of the eight live stream cameras, which to my mind is a simple, wonderful, modern, creative marketing tool. The rest of Sunday was spent resting as nobody got much sleep the night before on the bus.

In the evening I ventured out to a small local pub that was

holding a barbeque for some of the Harley Davison riders, feeling at first quite intimidated being surrounded by 20-plus leather clad bikers but after exchanging a few friendly words became more at ease and relaxed in their company. That night Michelle and I had the four-berth dorm to ourselves and we both enjoyed a good night's sleep.

Monday

Michelle was heading off today on her bike, the man from Japan was leaving on Tuesday and I planned to go to Ayres Rock on Wednesday. We all had breakfast together before saying goodbye to Michelle. I felt quite saddened as she pedalled down the road for I had developed both a liking and respect for her.

My first port of call after she had left was to a second-hand book shop because I wanted to start doing some reading and research into local indigenous Australian Aboriginals.

The main book I found and read was called *The Rainbow Spirit*; it was a short introduction to the Rainbow Spirit creation story and drew parallels with the creation story found in Genesis. In short it suggested that the Aboriginal concept of the Rainbow Spirit could easily be understood as their concept of God, in the sense that there was something greater and stronger than humans that had both created and interacted with the planet. My next stop was to be the Museum of Central Australia and Strehlow Research Centre.

Professor T.G.H. Strehlow had conducted an enormous amount of work and research into the local Arrernte

Aboriginal people. On the walk to the centre I passed two trees with a plaque by their side so I stopped to read it. Apparently, when the Stuart Highway was being built there were two trees directly in the path where the authorities wanted to put the road but these trees were of spiritual significance to the local Arrernte Aboriginal people because they were part of the rite of passage for their young men from boyhood to manhood. So what happened was that the elders of the Arrernte people agreed to the trees being cut down on the condition that two cuttings were taken from them and these cuttings then replanted as near as possible to the original site, with a plaque explaining the circumstances. What an excellent way to resolve a dispute.

During the walk at times I thought maybe I could hear the voices but was again unsure if they were real or imagined.

The Strehlow Centre itself was a little disappointing in that it told of the man and his works rather that giving details of his works themselves. However, I did discover that his main book, *Songs of Central Australia*, now out of print and very expensive to buy as a second-hand book, was available at the local library.

Outside the centre and under a tree was a large Aboriginal lady doing some painting, so I stopped and sat with her. Her name was Maggie. We sat and talked in each other's company for about ten minutes before I headed off to the Art Gallery. The gallery was featuring works of Aboriginal art from the young local artists and was fabulous. Then back to sit and drink tea with Maggie who was painting the tree that she was sitting under.

Her painting was from an aerial perspective; it then occurred to me that much of Aboriginal art is from this aerial viewpoint. Also, that maybe the dots that are used in this form of art could possibly be taken to represent the atoms we now know go to make up matter. Therefore Aboriginal art could be understood to be highly spiritually significant both in its perspective and its delivery. The fact that the Albert Namatjira Gallery is there totally escaped me, which as it turned out was a blessing in disguise.

Monday Night

It was tonight that the voices started again.

With Michelle gone I had the four-bed dorm to myself and as I lay my head on the pillow I wondered if anyone else would come into the room and, if so, what they may be like. Trying to sleep I started to notice odd sounds, very subtle at first, such as the quickening beat of the air conditioning fan or random noises from both inside and outside the dorm.

Then, "The wheels on the bus go round and round," said almost chanted or sung to the rhythm of the nursery rhyme. This continued for some time and now I was certain that the sounds were real and were not an audio hallucination. The only difference between the words being sung and the words in the nursery rhyme was that the words 'all day long' were replaced with the words 'all night long'. So they chanted, "The wheels on the bus go round and round, round and round, round and round, the wheels on the bus go round and round all NIGHT long."

The voices, their tone, the threatening nature of the wording, all combined to unnerve and frighten me as I tossed and turned in my bed, feeling increasingly unsettled. Then a woman's voice, the same voice that I had heard in Kashmir and India, clearly said, "What's your mission Andy Cole, what's your mission?"

Now different sounds started to happen, sounds that made me start to sweat; I could feel the goose bumps on my head, as small droplets of sweat started to form, my heart rate increased and my mind started racing.

Then a stern male voice one I had not heard before told me, "Our goal is world domination and if you come and join us you can be Governor of Gibraltar, if we cause any problems in achieving that, so be it. We will make recompense later."

Then the sounds, noises and voices continued at different sound levels, all short bursts and repeated over and over, musical sounds such as, "I'm Andy Cole but I'm not a soldier, I've got a soul but I'm not a soldier."

Short, sharp, clear bursts of sounds repeated over and over.

As the night went on I grew increasingly restless, increasingly concerned, increasingly frightened. I went downstairs to buy some water from the vending machine as my throat was becoming increasingly dry.

Returning to my room and trying to sleep the voices continued with various other noises and chants including, "Whether a Christian, Muslim, Hindu or a Jew then God loves them too."

Repeated, over and over, with different religions added and the order interchanged.

Then they told me that since they had last been in contact with me in Kashmir, they had been checking up on me and knew my whole life history and that this was easy using computer databases. They quoted some things which confirmed this.

The whole process went on for hours and at one point it seemed as if my thoughts were being read and repeated back to me. I was scared, nervous and uncertain of what was happening but at some point it must have stopped as I think I managed to get at least a little sleep that night.

Tuesday Daytime

I awoke early Tuesday morning and decided to change my plans and head off to Ayres Rock a day early, going directly to the Greyhound Bus Office to book my seat and confirm my ticket but unfortunately the office was closed and wasn't due to reopen until after the bus was due to leave.

Therefore I decided to return to the Backpackers, Melanka's, and started reading *Central Australian Religion-Personal Monototemism in a Polytotemic Community*; the title could be understood to mean the idea of one god in a multi-religious society, it was by T.G.H. Strehlow, and I had bought the book the day before at the research centre.

In a short, concise book, Strehlow outlines the basic principles he discovered about Aboriginal spiritual belief. It outlines the Aboriginal belief in the dualist nature of humankind, body and spirit; it deals with how spiritual belief permeates the whole of society and covers Aboriginal views on nature and how man and the planet are seen as part of the same living whole. He also argues that the battle between science and religion is responsible for much of today's psychological and spiritual malaise.

Inspired by his work I decided to head off to the library to read and study his greatest work, *Songs of Central Australia*.

On the way I called at the Greyhound Bus Office, which was now open, to book my ticket to Ayres Rock for the following day and wave goodbye to the man from Japan and for some reason I recall that as we said goodbye the song *I'm just a Boxer* by Simon and Garfunkel was playing on the coach radio.

It was while making my way to the library that I started to hear the voices again; trying to ignore them I continued to the library.

The receptionist told me that the book *Songs of Central Australia* was on the restricted list but I could read it in the study area. The book is a masterpiece; it deals with issues about the Aboriginal belief in the power of chanting, the power of poetry and the power of the full name. It deals with the link between chanting, music, dance and rituals and outlines Aboriginal belief in reincarnation and religion as magic. It speaks of the dream time, totems, the ritual used and the fact that death was not something that was feared. During most of my studies I could hear the voices.

Tiring of studying, reading and trying to block out the voices I decided to go for a walk along the dry river bed of the River Todd but the voices just continued. Then I tried to take refuge in some cafés which had music, in an attempt to block out the voices but it appeared that any songs that were playing only seemed to have some kind of connection to what the voices were saying to me; there seemed to be no escape from them. Moving from café to café I became increasingly unsettled, seeking refuge in the shopping centre but the sounds and music here also only added to my confusion and nervousness.

Eventually I went into the Aboriginal Art and Culture Centre; here at last I found some degree of peace and also, from a selection of only about 20 books, found *Aboriginal Men of High Degree* by A.P. Elkin, the book that had been recommended to me some weeks earlier by the receptionist at the Queensland Museum, during my first day out in Brisbane. This synchronistic event helped to settle and relax me.

In the evening I decided to have dinner at the Overlanders Steakhouse, which was a good choice; lots of Rolf Harris and Waltzing Matilda-type music which helped to block out any contact from the voices. After dinner I went to the Sounds of Starlight Theatre, which featured didgeridoos and other live musical instruments; here also the music and stories helped to block out the voices.

Tuesday Night

For the second night I had the four-bed dorm to myself; it was to be a long night. Armed with plenty of water and trying to sleep it was not long before the voices started with, "The wheels on the bus go round and round, you're going to walk on Holy ground, the wheels on the bus go round and round, you're going to walk on Holy ground."

Then they continued with a whole range of different noises, music and sounds, which as in the previous night made me increasingly nervous. A whole range of dialogue took place between us, so much so that I started to question if the voices had their origin inside my head or were from an external source; again, as during the previous night, it appeared as if my thoughts were being read and repeated back to me, which only added to my confusion.

It was at that stage that I heard people outside the bedroom window talking about the music; too frightened to get out of bed and investigate I listened intently. It appeared someone was complaining or at least asking questions about the music and where it was coming from; it sounded as if a police car drew up and there was then a discussion taking place regarding where the music was coming from.

Being too frightened, scared and nervous to get up and check, I stayed in my bed. That was until the voices told me that they wanted me to get up; so I got up and went to the toilet.

During the night my emotions were ever changing because sometimes the voices were friendly and encouraging and at other times frightening.

There was a chant of, "Come and join the boys, we'll get out all the toys," on the one hand and the threat of, "We going to have you shot at Ayres Rock," on the other.

Then there were chants of, "HIZ-bella, HIZ-bella, HIZ-bella," chanted by the male voice, over and over again,

"HIZ-bella, Hiz-bella, HIZ-bella."

Which added to my fear, added to my anxiety, added to my terror.

During this period the female voice spoke and she said in a mocking tone, "Of course he's terrifying you, he's a terrorist, he's only doing his job."

I tried self-hypnosis to try and get some sleep, I guess a form of meditation; this at some points offered some rest but not sleep.

At one stage they told me they wanted me to hire a car and drive west and that they also wanted me to place a bomb in Selfridges, Sydney when I got there and threatened that I would never leave Australia unless I did what they wanted.

They told me that I didn't need to die; they didn't need me to be a suicide bomber, that they just needed me to leave a bomb in Selfridges, Sydney.

My heart rate was increasing and I was aware of it pumping faster and faster, also my pulse rate was racing. Drinking plenty of water and visiting the toilet many times as the night went on I could smell the fear on my body. They knew what was happening to my body and seemed at some points to be able to monitor my heart and pulse rates, with comments such as, "Pop, that's another blood vessel gone."

They asked how much more of this I could take; five days was my reply.

They again told me their goal was world domination and if I joined them I could be Governor of Gibraltar and repeated the threat that if I did not join them, I would never leave Australia.

At times it seemed as if the voices and noises were filtered into the rhythm of the air-conditioning unit so that the rhythmic sound of which could be interpreted as words.

At one stage I recalled something I had read that the Dalai Lama had said about how our enemies offered opportunity for personal growth. This offered some comfort for a while.

It was about 4am that they told me to get my mobile phone and ring a phone number, which they gave me and I wrote down, but because I had never changed the sim card in my phone since I had been in Kashmir, my phone would not work. They then made me promise to get a new sim card first thing in the morning and with that told me to go outside to make the call.

So I got up and went downstairs to make the call from the phone boxes just outside Melanka's. Going outside, it was dark, the street was deserted, and my hands were trembling because I was so nervous and so frightened; because of this I couldn't make the phone work. Then a taxi drew up about a hundred metres down the street, it appeared to flash its rear lights at me before pulling into a side street. I considered following it but good sense or fear prevailed and I stayed by the phone boxes.

The taxi then must have turned around and came driving slowly towards me. I held out my hand to stop it and talk to the driver and asked if he was looking for me. He said he wasn't and we said goodnight. I went back to my room and tried to rest, having been unable to make the phone call; it would be dawn soon and there would be people about and those two things would at least make me feel psychologically safer.

Wednesday

It must have been between 5am and 6am that I went downstairs to reception and helped myself to the free tea and toast that was on offer. There were backpackers checking in and some checking out, some going on tours and others like myself just resting. This offered some respite from the traumas of the previous night.

Tired but relaxing I started reading A.P. Elkin's book, *Aboriginal Men of High Degree*. He spoke of the Aboriginal rites of passage and how this involves facing innermost fears; the book also discusses numerous parapsychology events that the indigenous Aboriginal can experience, such as telekinesis, out of body experiences and telepathy. On the subject of telepathy, he states, "The telepathy that Aboriginal men of high degree practise is also practised by the Tibetan psychic adepts. The Tibetans assert that telepathy is a science..."

The book offered a haven to retreat into but at times the music and even the adverts from the radio frightened and unnerved me as they appeared to echo some of the themes from the previous night with songs such as 'Those nights of fright and those wonderful sights'.

During the morning I went out to get a new sim card as I had promised to do the previous night and while doing so decided to postpone my trip to Ayres Rock until Friday, being too frightened, too nervous and too tired to make the trip today.

It was about mid-morning that the voices told me to take my mobile phone and walk away from Melanka's to somewhere quiet and wait for a phone call. I walked to a park just down the road from Melanka's, found a quiet spot under a tree in the middle of the park and sat and waited.

As I waited a soft breeze slowly moved the branches of the tree I was sitting under and at times allowed the sunlight to gently filter through to my face. Sitting there I felt very peaceful, almost serene, and at one with my God, calm, relaxed and at peace with the breeze, the sunlight and the sounds of nature.

It was in this relaxed state of mind that it occurred to me that if the voices had been repeating my thoughts back to me, then they must have been reading my thoughts and, as I could hear them, this meant that we were communicating with each other – mentally.

With this I felt a great sense of joy and engaged in a long conversation with them. The conversation took place mentally, they could read my thoughts; I considered this must be being done by telepathy and I could hear them but was unsure by what means. We discussed how this type of conversation might work and I suggested, recalling the article that I had read in the *New Scientist*, that maybe it was possible because of the work being done in the field of the Quantum Theory of Entanglement, where what is done to one proton affects another proton, instantly.

During our conversation we had a disagreement about the music the previous night and because of this I went to the police station to enquire if there had been any reports filed regarding the music. They were unhappy about this visit to the police station and again started saying things that threatened me. I took refuge in the library and continued reading Strehlow's book.

On my way back from the library to Melanka's there was a group of four Harley Davidson riders riding down the road; the voices told me that the riders had come to kidnap me and that it would be easy for them to do so. Scared, I waited as the four riders drove slowly past me at the very spot I had waited while the taxi drove past the previous night. The riders went into Bojangles and either out of foolishness or as a show of bravery I followed them in.

While in Bojangles I got into conversation with a man called Patrick, a man about my age. This conversation helped to settle me and calm my nerves and also resulted in the voices saying that there would be another time for them to kidnap me, as the bikers left.

After the bikers had left I headed back to the safety of the Backpackers after arranging to meet my new friend Patrick in the bar there at 6pm.

Wednesday Evening

While I waited for Patrick I sat and thought about the events and experiences of the previous few days. There were definitely two main voices, one male, and one female. They were reading my thoughts instantaneously, this I was increasingly sure being done by telepathy and I could hear them. These things I was certain of. At times I had been extremely frightened and at times calm plus the whole range of emotions in between.

Patrick arrived and it appeared that he had just ended a relationship in Darwin and was now moving to Alice Springs in search of work as a teacher. He told me of some of his teaching posts in the outback and how he would need to fly by light aircraft to reach the settlements.

As we sat and talked, almost unnoticed a number of vehicles started drawing up outside the bar. The female voice told me that I was to face a kangaroo court. She said the reason for this was some of the local drug dealers were unhappy about me going to the police station and that they needed to be convinced that I was not a threat to their livelihoods. She asked if I would take a glass in the face.

She told me to approach a man in the bar and as I did so he asked me if I wanted a game of pool and with that took me to the pool table. He asked me to crouch down and look at where the coin mechanism should have been but the mechanism had been removed and had left a large space big enough to hold a glass. Squatting down next to the table unsure of what might happen I waited but all that happened was the man showed me how to get some free games of pool. There then followed some games, firstly between the man and myself and then between me and Patrick.

During these games of pool the female voice was very encouraging and calming, teasing me about my lack of pool skills. She said I need to remain calm in order to gain the respect of the dealers involved in the kangaroo court.

Eventually Patrick went to bed and I sat and waited for the court to take place, noticing that the back exit was guarded by a large truck and the front door, which was half closed and had a number of people standing there, thus blocking any attempt to escape. I sat and waited while considering a whole range of things that they may do to me. Alice Springs is 2,500 kilometres from any major town and I believed anything was possible.

Eventually the male voice told me to get on the stage and dance with the others; I did so, unsure of the dance moves that everybody else seemed to know, moving but hardly dancing with the crowd. As the music finished everyone left the stage and I was left there alone. I stood and waited for what would happen next.

After a short time the female voice told me to sit down and wait. I did so.

About ten minutes passed before the male voice told me it was time to get back on the stage. I did so.

Waiting, alone on the stage for some time, unsure of what was expected of me, before I eventually approached the DJ and asked what it was that I should do; her answer was, "Dance if you want to dance, sit if you want to sit."

So I sat down, almost trembling with fear. The female voice now taunted me with, "Where's your God now, where's your God now?"

"Deep inside," was my reply.

A further ten minutes or so passed, before the female

voice said, "Fuck off, you're free to go."

Exhilarated and relieved I left the bar and went outside to calm down and relax before going to bed.

Wednesday Night

For the third night in a row I had the four-bed dorm to myself but for the third night in a row sleep was going to prove difficult. The noises and voices continued in much the same way as the previous nights but with the added pressure of the sounds of the Harley Davidson bikes riding past the window and the voices informing me that these were the riders from earlier in the day and that tonight they were going to come to my room and kidnap me. At different times different threats were made, such as the threat of an alligator being thrown into my room.

As with the previous nights my emotions vacillated from one extreme to another.

During the night I considered how the telepathy might be working and thought that it may be being done via satellite.

"No," was their reply, "it was being done by Sputnik."

At some points during the night I felt I was being psychologically traumatised but at the same time felt a certain fondness of those responsible.

In many ways this was an easier night for me because either through exhaustion or relief after the kangaroo court, some sleep was possible.

Thursday

Waking early and feeling somewhat refreshed I made my way downstairs for tea and toast; Patrick soon joined me, as did a girl who was reading *The Alchemist*. It was a book I had been aware of but never read; Alchemy was the sixteenth century art of turning base metal into gold. Making a mental note to read it I considered that it could be seen as an analogy of turning the base human condition into a spiritual one.

Because I was feeling refreshed, I decided to hire a car and explore and asked Patrick if he wanted to join me, which he did. We hired a Ute and set off on our journey; the first planned stop was to be Standley Chasm about 50 kilometres away. As we drove the voices continued but either because I was in company, or because I was getting used to them, or because for all the threats that they had made, nothing had happened, I was growing increasingly comfortable about them and as we drove along the long straight road of the West MacDonnell Range, I felt a certain sense of freedom and liberation, singing to myself and sticking two fingers in the air to the satellite.

The voices had been with me in Kashmir and India and had now followed me to Australia. What was it all about? I didn't know but for now I was comfortable and growing increasingly confident about the situation.

We explored Standley Chasm, with its narrow gorge, high sided cliffs and its many flies before heading on to Hermannsburg, which is a restricted Aboriginal settlement a further 75 kilometres away. The road was long and straight with not a building in sight, totally unspoiled and now singing along with the voices, "Life is but a journey and where that journey goes..." followed by words that now escape me.

Because the town is restricted there are some areas that visitors can't go but most of the town was open to us and no permits were required. There was a shop which had a list of some basic rules of the community displayed. These included no alcohol, no drugs, no sniffing of petrol along with the penalties incurred for breaking the community's laws. It struck me, recalling the licensing laws of Tennant Creek, how different communities had different rules and laws and how what is acceptable in one community or culture is unacceptable or unlawful in another.

Whist exploring we found the nineteenth century Lutheran church; there was a bell at the front of the church and a mortuary at the rear. It occurred to me that the bell would have been used to call the indigenous Aboriginal people to church so that they may learn about God but, in doing so, how the Church would have been calling them away from their own culture's natural spirituality. Although the Church's motivation may have been good they seemed to be taking something away from the indigenous people, something that was both natural and instinctive.

In death too the Church seemed to have taken something away, being laid to rest on a concrete slab in a concrete building rather than just being returned to nature in a natural way, and I wondered if the Aboriginal had really gained anything from their conversion to Christianity.

Across the courtyard from the church was a small art gallery and what an experience this proved to be! It was featuring a display of the work of Albert Namatjira; this was the perfect setting in which to discover the man and his work. Albert Namatjira was a genius; he had taken the European style of painting landscapes and painted them from an indigenous perspective. So that those things in the landscape, trees, rocks, cloud, all things could be seen to have faces or hands or some other body parts or, put another way, he had painted the spirit of the land living in the land. The spirit of the land was a living thing, something real, something seen, something experienced, an example of a living God. An expression of people's spirituality expressed in art but more than that, expressed in the conquering culture's art form, brilliant.

Unfortunately Albert Namatjira's life became complex. He became the first Aboriginal to become an Australian citizen, meaning he was allowed to buy alcohol but, being true to his roots, he would supply it to other local Aborigines who were not allowed to buy it. For doing this he was jailed, dying soon after his release.

Patrick told me that Archie Roach, an Australian singer, had written a song about Albert Namatjira and if there was one thing I took back from Australia it should be this song.

We were hot and dry and because Hermannsburg was a dry town and the next town was an hundred kilometres in the opposite direction we decided to return to Alice Springs. As we drove back the West MacDonnell Range was glistening in the sunlight and you could clearly see faces in the mountains, faces and objects that would only reveal themselves if the sun hit them in a certain way or if a shadow caught them just right. Images that were real and easily accessible but only accessible in a certain light, each manifestation requiring a different degree of light or shadow to enable it to reveal itself, the planet a living thing indeed.

We arrived back in Alice Springs and decided to have a beer at the local RNLI club. It was here that the voices started to be aggressive again. This time, in addition to the voices, I experienced sensations of hot and cold so that at one moment it would seem that there was an ice cube on my hand then the next felt like it was being burned. The Pokies machine starting making noises that seemed to be taunting, "You're going cuckoo, you're going cuckoo."

Trying to ignore them I continued in conversation with Patrick before we both headed to Bonjangles for one last beer. While there a power cut plunged the bar into darkness. Now these events started to unnerve and frighten me again so I left, arranging to meet Patrick at Melanka's bar later.

Thursday Night and Friday

As I sat and waited for Patrick the voices started to question me about Marian. I told them the story of the Polish man in Gibraltar who had asked for a ticket home to Poland, some months earlier. Until that point I hadn't really considered that there may have been a connection between the voices and the Polish man, Marian.

Patrick arrived. We sat, chatted and drank; he told me he was broke and that I could have sex with him if I wanted. He said he could only afford to pay for another few nights at the Backpackers and then he was on the streets. He asked if I would pay his fare back to Darwin or if I would pay for some accommodation for him. I said that I wouldn't give him cash but that I would pay his accommodation for the next two weeks, which would give him the time to find himself a job but declined the offer of sex. He was more than happy with this and we talked some more, for my part I reasoned that his company and conversation was a distraction from the voices.

I had a fairly early night and packed my backpack ready for the trip to Ayres Rock the following day. For the fourth night I had the four-bed dorm to myself.

Tonight a lot of mental dialogue would take place; the difference was that tonight I felt stronger and less frightened of the situation. Again the voices threatened me that they were going to have me met and shot at Ayres Rock but I was determined that they would not scare me from making the trip as they had scared me from making the trip to meet the Sufi Baba in Kashmir.

They reminded me of the instruction they had given me two nights ago, firstly to take a car and drive west and secondly to put a bomb in Selfridges, Sydney. They reminded me that I had hired a car and driven west and now they wanted the second part of the instruction carried out and again repeated the threat that if I didn't join them, didn't do what they wanted, I would never leave Australia.

It must have been about 4am when, half asleep, I suddenly felt a sharp sensation in my chest, as if something had hit me. Pulling myself to the side of the bed and half getting up, I must have felt a second sensation because the next thing I knew was I was lying on the floor.

Staggering to my feet only to collapse onto the floor I must have banged my head as I fell and this must have knocked me unconscious for a short while.

I'm unsure for how long I was unconscious, or what had happened, it felt as if something had hit me or maybe I had had a heart attack, coming round and noticing blood on the floor and a cut on my head. Feeling totally weakened, totally drained and totally confused, I staggered to my feet and slowly made my way to the toilet.

My bowels exploded in a cascade of excrement and my

torso exploded in a shower of sweat. Half collapsed, not knowing what had happened, I must have sat there for almost an hour while my body slowly regained some strength, unsure if I was dying or if I was having a heart attack or if I had been stunned by something.

Eventually, I slowly and gingerly made my way back to my room and there on the ceiling was a small white laser dot. I immediately thought that it was part of a beam that had been used to stun me and put the light on as a form of defence.

The female voice said, "That was for my mother."

I must have insulted her mother during the night's dialogue but at least her comments confirmed what had happened.

Weakened and confused I made my way downstairs and outside; there the fresh air and hot sweet tea would help me to recover a little more.

As I sat and rested I could slowly feel my strength coming back to me and remembered the ticket I had booked to Ayres Rock and considered still making the trip but the female voice told me, "Don't be stupid, get yourself to the bloody hospital; you've just had a heart attack."

I considered this good advice, so that's what I did.

I made my way on foot to the hospital about 300 metres down the road from Melanka's, registered in the accident and emergency department, then went outside for a cigarette.

It was while pacing about outside the hospital that something extraordinary happened. While pacing I was stopped in my tracks as both my feet straddled what can only be described as an Arabian sword. The sword was drawn on the pavement in what seemed like blue paint or blue chalk. It had a highly decorated handle and a long curved blade.

The female voice said, "Who is this man; I must get a picture of this."

I stood there, looking down at my feet either side of the sword in amazement. Given the journey I was on, here in Alice Springs, miles from anywhere, outside the accident and emergency department with my feet straddling the sword, it had to be a sign of some kind. The female voice informed me that the sword was the sword of Jihad.

Shortly afterwards I was called in to see the doctor. She was a young, very beautiful Indian doctor and as we spoke I discovered her family were from Kashmir. She examined me thoroughly, checking my heart rate, pulse rate and blood pressure before reassuring me that I hadn't had a heart attack and my heath seemed fine; she then dressed the cut to my head. During the examination I told her of the events of the previous four nights and we also spoke of my experiences in Kashmir. After the examination and conversation she transferred me to a side ward to await further attention.

I must have stayed in the side ward for a few hours before another doctor, also Indian but this time male and wearing an England cricket shirt came in and arranged for me to have a CAT brain scan. All of my results came back fine but the male doctor suggested that they would like to keep me in for a few days for general observation, if I agreed; I readily did so. With that I was admitted to the mental health wing of Alice Springs Hospital.

The unit itself was clear, bright and comfortable with three other patients who were all friendly. To the rear of the ward was a walled garden area with a mural of the outback painted on the wall and a games room with a CD player.

During the evening I started hearing the voices again and

the male voice told me that I still wasn't safe and that they could still have me killed while I was in the hospital. It was becoming clear that the female voice was increasingly friendly and encouraging while the male voice continued to threaten, frighten and scare me. It was the male voice who told me during the evening that they had kidnapped and killed my family; this of course intensified my fear. My sleep that night was aided by two Temazepam tablets and I had the first full night's sleep since leaving Darwin over a week ago.

The Weekend

The first thing I did on waking was to ring my family; it was the weekend of my daughter Michelle's birthday; to ensure they were alright. To my deep relief everything was fine in England. I didn't tell them I was in the mental health unit of Alice Springs Hospital, not wanting to worry them after the phone call from the British High Commission in India.

With that done I felt a great sense of release of fear and, having slept well, felt rested and, feeling secure in my environment, was able to think and contemplate on what had happened during the previous week.

Who were these voices? What did they want from me?

The day was spent resting, relaxing, talking to the other patients and thinking. I thought about religion, its symbolism, its meaning and its interpretation and pondered that by following my path I had arrived here; what was it that my God wanted me to do? What was it that I should do for my God? Thoughts of the sword of Jihad and the laser stun gun drifted around my head.

It was early evening when the male voice told me that the stun gun they had used on me was on a very low setting and that they now needed to kill me because I knew too much and that I was to shortly go outside to face them. This again raised the question for me about would and could I face death, would I again be prepared to face death for my God. During the evening I again became increasingly nervous and my hands started to sweat as I thought about what might lie ahead.

The time came and I went outside and faced the direction they told me to face; there in the sky was a satellite, quite low and clearly visible, it also seemed very bright. I stood and looked at it, waiting for what would happen; if the stun gun had been on a low setting and if they were able to increase its power then I believed they would have been able to kill me. I stood and waited for what would happen!

For whatever reason they decided not to harm me but rather asked me to join them in their cause; they said there were many things I could do for them, for example share dealing, because sometimes share prices go up, and with that the satellite moved up. And sometimes share prices go down and with that the satellite moved down. Then the satellite moved from left to right, then all over the place. Then they said oil prices go up as the satellite moved up again and oil prices go down as the satellite moved down again. I stood there in total awe, in absolute amazement.

They were real, they were absolutely real, the voices were real and they had been reading my thoughts and communicating with me via satellite. I was so happy, so delighted, such a great feeling of release. It seemed that they now trusted me and also now I trusted them. The male voice reminded me that it wasn't a satellite but a sputnik, while the female voice told me I was very lucky because for tonight at least I had a satellite, or sputnik, all to myself. The feelings of joy, the feelings of release, were intense.

I went back inside the unit but told no one of my experiences. After a short time they told me that I had to go outside and face them again, which I found a lot easier than the first time. They could have killed me but didn't, it was a big night, we now seemed to trust each other, seemed to have bonded. My sleep that night was again aided by Temazepam.

Sunday was peaceful and the day was spent relaxing and talking to the other inmates. One of them, a local man, explained the mural on the garden wall. It showed the bush after the Aboriginals had burnt it, to clear it and the first plants that grow back were like little weeds but were in fact marijuana plants, drugs but natural drugs. He also explained how these plants or drugs were used in some Aboriginal rituals.

Also during the day I spent time thinking about personal religious experiences and recalled books by R. Otto and W. James that I had read many years ago. I came to the position that personal religious experience may call a person to act on those experiences, to act on their beliefs, which was the point W. James was making when he wrote, "By their fruits you shall know them". I had followed my coincidences, followed the synchronistic events outlined in this book and this is where it had brought me.

It was Sunday evening when another surprising thing happened; Patrick checked into the unit. It seemed that he had cashed in the money I had paid for his room and gone out on a drunken spree and in a state of drunken depression had been admitted to the unit. We were both surprised to see each other and I was angry when I discovered what he had done and demanded back any cash he had left, which may have been harsh.

The next morning would be Monday and that was when all the inmates had to face the panel of psychiatrists and doctors to assess who should stay and who should be discharged.

Later that evening, the voices gave me some instructions and advice for the meeting with the panel of psychiatrists and doctors. They told me that the psychiatrists may offer me two white tablets and if I took the two white tablets then the voices would end and I wouldn't hear them anymore, it would be over.

Secondly, they said that if the psychiatrists asked who the voices were, whatever I did not to mention Al-Qaeda and that I needed to be very clear on that point, not to mention Al-Qaeda.

Monday

Waking early the following morning I waited in the games room until it was my turn to see the panel of psychiatrists and doctors. While waiting I wanted to listen to some music so went to put on the CD player. There, unbelievably, on the CD player was the album by Archie Roach featuring the song about Albert Namatjira, which Patrick had recommended to me on our day out in Hermannsburg. Absolutely surreal, for me absolute confirmation that I was still on my right path.

The voices repeated their instructions from the night before. That if I was offered two white tablets and took them then I wouldn't hear the voices anymore and secondly, and most importantly, not to mention Al-Qaeda.

There were about ten psychiatrists and doctors as I entered the room, two of them looked particularly heavyweight; we discussed my situation.

I told them my theory about Carl Jung, of whom of course they were well aware. I told them that I believed Jung to have been totally correct with his concept about the collective subconscious and his concept of synchronicity. I expanded the conversation about the collective subconscious to show I was aware of the idea of us all being connected via the subconscious and discussed details of how I had been following the synchronistic events on my journey. I also told them about the voices that I had been hearing and how these voices were now part of that journey.

They then asked me, "Who do you think is responsible for these voices, who do you think they are?"

I replied that I wouldn't like to speculate.

They then asked me, "Are you happy to continue on your journey with the voices, to see where it takes you?"

I replied that I was more than happy to continue, to discover where it would take me.

They then asked me my plans; I told them that I intended to visit Ayres Rock before heading on to Coober Pedy, then Adelaide, Melbourne and finally Sydney, to see my sister Sheila. They reconfirmed that I was happy to go on with the voices and reconfirmed that I would be meeting family later in my trip.

With that they discharged me.

(The relevance of that meeting didn't dawn on me for over a year.)

As I walked out of the interview room the voices said, "We like your style Andrew Cole, we like your style, I wouldn't like to speculate, not only didn't you mention Al-Qaeda but you said, I wouldn't like to speculate, we like your style."

With a spring in my step I left the hospital, rebooked into Melanka's for one last night and rebooked my ticket to Ayres Rock for the following day.

The rest of the day was peaceful and spent relaxing and searching for books on world religions. The first book I found was edited by W. Clarke and featured a chapter on Buddhism by Peggy Morgan, who had been one of my lecturers at Westminster College, Oxford, and the lecturer who had housed the Alistair Hardy Centre in her garage, the second book was by W. Owen Cole. Both books offered an overview of the major religions and as Islam was the only major religion that I hadn't studied at university, combined with my experience of the Arabian sword, or sword of Jihad, outside Alice Springs Hospital, I decided to read the sections on Islam first and planned to do so next day, on my journey to Ayres Rock.

That night was peaceful and because the voices and I continued to communicate, I decided to give them a name, feeling now that I had proof that they were real and not in any way a figment of my imagination; the proof of their existence for me was seeing the satellite moving combined with the laser dot on the ceiling.

I decided to call the female voice Gita, after the Hindu Holy book *The Bhagavad Gita*. I didn't name the male voice but he told me he was a Tottenham fan; it was unclear if he meant Tottenham Hotspurs or the Tottenham Ayatollah, in any event he became Tottenham man.

From this point on the voices would become my constant companions.

Off to Ayers Rock

Finally I boarded the Greyhound coach to Ayers Rock, at the third attempt. We set off on the five and a half hour trip. Our first stop was about an hour down the Track at the Frontier Camel Farm, where I played with a pet dingo and surveyed the wildlife. We then re-boarded the coach and I got engrossed in reading my new books on world religions.

I started reading about Islam and the five pillars of faith. A belief in one God, prayer as a way of worshiping God, fasting as a means of self-purification and understanding the suffering of others, compassion, which means caring for and giving to others, and pilgrimage.

As I had seen the Arabian sword, directly between my feet, painted onto the pavement outside the hospital I was particularly interested in reading about Jihad and learnt that the word is often misunderstood in the West. There are two types of Jihad, the inner Jihad and the outer Jihad, the inner Jihad being considered the more important and that this could best be understood to mean an inner spiritual struggle or spiritual journey.

The book also gave an overview of the history of Islam, the proof of Islam and in particular the Sufi tradition; it wetted my appetite for further study.

The journey soon passed and we arrived in Yulara, the resort for Ayers Rock and pulled up outside the Outback Pioneer Lodge. I booked two nights in a shared four-bed dorm, which was the only accommodation left. The cost of the room and also the cost of refreshments were a lot more expensive here than in Alice Springs or Darwin. Gita, the female voice, suggested that I go to the local shop to buy some supplies because, that way, my stay would work out a lot cheaper. This was again good advice and with that done I booked my ticket for the tour of the Rock for the following morning, booked a ticket for the Night Sky Show that night and booked a ticket for Kata Tjuta, for the morning after.

With all my arrangements now in place I made my way to the central viewing area in Yulara to watch the sunset. It was a magnificent experience, with Uluru (Ayers Rock) to the left and Kata Tjuta (The Olgas) directly in front, as the sun set slightly to the right. The huge rock, seen in so many photographs and pictures but to be there and experience its vast natural beauty in its unspoiled natural environment was for me deeply spiritually touching, staying until the sun had sunk beyond the horizon, just drinking in the atmosphere.

That evening I cooked myself some pasta in the Outback Pioneer's excellent communal kitchen before sharing the four-bed dorm with three strangers.

I woke early and boarded the bus for the short ride from Yulara to Ayers Rock. Initially we stayed in a group to watch the sunrise but the true impact of the Rock didn't really hit me until I was on my own. Just me and the Rock, oblivious to other people around I just stood in absolute awe, as tears streamed down my face; it was truly mesmerising.

The sun's rays were ever changing as they danced over the Rock in a kaleidoscope of colour. The natural ingrained patterns that nature had left over the centuries, which became clear and then disappeared as both the sun's rays and shadows combined to enhance the Rock's natural beauty. It was a moment of moments for me for a number of reasons. Firstly, the natural awe of the Rock and its setting were overwhelming; secondly, its natural beauty is magnificent; and thirdly, it was a long-held personal ambition fulfilled, it is a totally spiritual place.

To the right of the Rock is a pathway that was used by tourists to climb but the local Anangu Aboriginal ask that tourists don't climb the Rock because it is an extremely important site and of great spiritual significance to them, being part of the rite of passage of their ancestors. However, a great many tourists do climb the Rock, so I asked a warden why they allowed this climbing to take place if it was against the land owners' wishes. He told me it was a commercial decision taken by the Anangu Aboriginal elders and that they had the power to stop people climbing should they so wish; it seemed that the Anangu elders were torn between that which is spiritual and that which is commercial, rather like our world today.

After that I set off on the Mala Walk but with no intention of completing it, I just wanted to enjoy it, in my own time, at my own pace and in my own way.

All too quickly the tour was over and it was time to re-board the bus back to Yulara centre. As we drove away I couldn't help keep looking back, just to enjoy it, to feel it, to sample it, to immerse myself in it, for a little while longer and I did so with tears trickling down my face.

We returned to the Pioneer Lodge at Yulara and I sat and drank lots of Earl Grey tea and thought about my experiences. No one had turned up to shoot me as the voices had threatened and I had fulfilled a long-held spiritual ambition; the events and the journey had been worth it.

That evening I had planned to treat myself to a good meal in the Pioneer Barbecue, where you barbecue your own dinner, listen to the live music then head off to the Night Sky Show, to enjoy the stars, which was all fine until the voices told me that they had arranged to book the other three beds in the four-bed dorm but that only one person would check in and that he was coming with the express intention of killing me.

With that I returned to the dorm to check out what they had said to me. Indeed the three people from the previous night had checked out and in their place was just one person. He was mid- to late-thirties, looked in very good shape and was English. I asked him what he did for a living; he said he was a policeman in Bosnia.

A little nervous and confused I returned to the Pioneer Barbecue area to cook my dinner. As I did so my new roommate followed me and we got into conversation. I told him that I had started my journey in southern Spain and he questioned me about the Alhambra and Cordoba, both important Islamic sites. His presence and his questions increased my levels of anxiety, so much so that directly after dinner I went straight to bed, and tried to rest. However, this proved difficult, as thoughts of what the voices had threatened ran through my mind. They had been correct, there was only one other person in the four-bed dorm.

It was with increasing tension levels that I decided to miss the Night Sky Show and stayed in bed, scared; sleep was impossible even with the aid of Temazepam. It must have been about three in the morning, when the door of the dorm opened and in walked my new roommate. I waited, exposed my throat, in an action signifying my willingness to die but nothing happened. He went to bed and I must have dropped off to sleep, relieved, soon after. It seemed that I was increasingly able to cope with my fears.

By the time I woke up he was gone and I took the tour to Kata Tjuta, which is a fragment collection of rocks and very impressive. On arrival I set off on the walk of the Valley of the

Winds but, as with yesterday's Mala Walk, didn't want to complete it, rather I just wanted to enjoy being there. The afternoon was spent at Walpa Gorge where the voices instructed me to prostate myself in act of worship; this I did, although feeling uncomfortable about it.

The final visit of the day was to a viewing area that enables spectacular views of Uluru, Kata Tjuta and unsoiled lands, as far as the eye could see. Nature, at its natural best, in its natural element with no man-made distractions; you could just imagine herds of wild animals roaming across the open plains with the two sets of spectacular rock formations in the background, beautiful.

All too soon it was time to return to Yulara and board the Greyhound bus back to Alice Springs before continuing the journey.

Off to Coober Pedy

As there is no Greyhound service heading south from Yulara it was necessary to return to Alice Springs for one last night before travelling south to Coober Pedy. Following my experiences at the Pioneer Lodge I was done with shared dorms and booked a double room for myself.

The trip from Alice Springs to Coober Pedy is about eight hours and I spent the time reading, studying, thinking, relaxing and communicating with the voices. Because a large part of the town is underground I wondered if I would still be able to hear the voices when we arrived there. I also wondered if I would see Dutch Michelle on the way because that was where she was heading.

The first impression of the town was, as my guide book had suggested, like something out of the Wild West. Apparently the police station, newspaper office and courthouse had all been bombed over the years, and something of a gold rush mentality still existed in the town.

On arrival I checked into the Radeka's Downunder Dugout Motel, which offered good accommodation at a reasonable rate. The double room I booked was six foot underground; they also had shared accommodation in an alcove area which was about 20 foot below ground. Unfortunately, I discovered that, however the voices were working, they could still function underground; I could still hear them and they could still read my thoughts.

I booked in for one night only, before finding myself an underground restaurant to have dinner in and, while there, I shared a table with an American woman, the only other diner in the restaurant. The conversation revolved around those things that were in my mind at that time, religion, Islam and travel.

The following day the voices suggested that I should fast for the day, at least not eat as a way of expressing my faith, which I agreed to do and then set off to explore the church next door to Radeka's before going in search of the Serbian Orthodox Church, which was an underground church, situated about two kilometres out of town in the desert.

While walking out into the desert I hailed a motorist to ask directions. He introduced himself as Richard Khan and said he would be happy to give me a lift to the church. On the way he told me that he was a local opal miner and poet and that his family lived in Sydney because it was better for his children's education. He also told me he was a Muslim and gave me the name and address of his imam and his mosque, Bali el Huda, in Sydney and invited me to visit when I reached there.

I thanked him for his hospitality, conversation and the lift and then set about exploring the church. As I was doing so the priest came and introduced himself and invited me to join him for a pot of tea. We discussed faith and how faith needs an expression, community and how community helps to build and strengthen faith, along with other interesting theological points. His parting gift to me was a book, *Against False Union*. Taking the gift I started out on the hot, slow walk back to Radeka's to begin reading it.

It was while reading it that another memorable event happened. I had just read the passage, "Thus, at the time when more than ever the West needed the spiritual assistance and guidance of the East, the chasm appeared, terrifying to the eyes of all," when an enormous wind suddenly whipped up. It was so strong and sudden that the dust swirled in the air and dead bracken was sent flying everywhere. It had such an effect on me that I jumped to my feet, threw my arms in the air and shouted, Allah Akbar!

I stood and considered what I had just read and the event that had just happened at exactly that moment and fully understood the coincidence.

With thoughts of the sudden wind, flying bracken, swirling dust and the passage I had just read running through my mind I went for a short walk along the main street, but the opal stores held no interest for me as I waited to pass time before boarding the Greyhound bus to Adelaide.

Off to Adelaide

The coach trip from Coober Pedy to Adelaide was a little less than 900 kilometres and would take about ten hours and because we left as the sun was setting this meant that the journey would be through the night. This offered an ideal opportunity to think and reflect on my situation and the events that had happened; the area that most concerned me was how did this telepathy work?

All I could think of was that thoughts are energy, thoughts are energy. Gita, the female voice, kept on confirming my thinking, "Yes Andrew, thoughts are energy."

The statement kept running through my mind, thoughts are energy, thoughts are energy and if this is the case then this would mean that the thoughts we have are brainwave activity. These brainwave activities give off a small but significant electrical charge. It was this small electrical charge that was being both intercepted and interpreted or, simply put, the energy from my thoughts was being remotely read.

Although unclear as to how this worked as a scientific formula, I knew that this is what was happening.

The sounds and voices that I had been hearing operated, in my view, differently. These I believed were achieved by use of a high frequency speaker system but I felt a lot less certain regarding this part of the experiences.

However, I was clear that there were two aspects to the experience, one that intercepted, interpreted and translated my thoughts and the other that allowed me to hear the voices.

We travelled through the night, passing the once top

secret Woomera Air Base; it was during the middle part of the twentieth century that science had learnt to split the atom and created the nuclear bombs that used to be housed here. Now it seemed that science had learnt to split the neuron or at least interpret the neurons, if that was how this technology was working.

Occasionally along the route we would stop to pick up or drop off mail and I would take the opportunity to leave the coach and enjoy the night sky. The stars, of which there now appeared so many, shone so brightly and clearly here in the desert, with no glare from artificial lights to mask their beauty.

It was the middle of the night when we pulled into Port Augusta for a midnight feast; I was starving, having fasted the day before. It was a place that I would have liked to stay longer, had I planned the trip differently.

The remaining part of the trip was spent resting and trying to sleep but I had a nagging doubt that I had missed something important in reviewing the events of the past two weeks.

We arrived in Adelaide early morning and I booked myself into a modest hotel for one night, then, because of tiredness and mental fatigue, slept until midday.

On waking I set about exploring the city, making my way first to the river for lunch and then, following the sound of music, over the bridge to the football stadium where a testimonial veterans' game was taking place between The Crows and The Power, two local teams. The wrist band which I bought as an entry ticket had written on it the word, 'Believe'! This made me smile inwardly as I said hello to my God.

It was the first AFL game I had been to and therefore

didn't understand the rules but two young ladies in the seats in front of me not only shared their wine with me but also explained the rules and finer points of the game.

The voices however started making me a little nervous by telling me that they thought that I was being watched and followed. After the game the voices told me to buy a newspaper; it had been the weekend of the bombing in Bali. They told me that my reaction to the photographs of the bomb victims in the paper would tell them if I was really on their side or not. They said they saw the bombing as a positive action. Feeling torn, confused, alone and trapped; being told by the voices that I was being watched and followed, I went back to the hotel and to bed, to escape the situation.

There the voices sang me to sleep with the song: "I've been through the desert on the horse with no name, it does you good to stay out of the rain, in the desert you can remember your name for there ain't no one for to give you no pain..." and, with the aid of the Temazepam, slept soundly till morning.

The next morning I went out to a local open-air café for tea and breakfast and as I did so a 4x4 drew up on the opposite side of the street. The driver stayed in his car; the voices told me that the driver was watching me. Then two men came and sat on the table next to me; the voices told me that the men were from Australian Intelligence.

Gita, the female voice, told me that I had to now be very careful, as the final part of any operation was always the most dangerous and that she was now certain that I was being followed. Her instructions were for me to behave just like a tourist.

Therefore, I took one of the city's bus tours, stopping at

the Art Gallery of Southern Australia and the South Australia Museum but found both venues uninspiring, which, given the circumstances, was perhaps not surprising. The time could not pass fast enough as I wanted to leave Adelaide as soon as possible, feeling very uncomfortable about my experiences while there. That evening as I boarded the Greyhound bus to Melbourne, I watched as a sniffer dog checked my rucksack.

Off to Melbourne

The journey from Adelaide to Melbourne would also be through the night. The voices told me that they would send one of their people on the bus to watch me. Looking around the bus as we set off, trying to identify who it may be, there was a young hooded Asian man sat across the aisle from me. We looked at each other and smiled, maybe it was him but of course I couldn't be certain.

As we drove I considered possible uses for this type of mind reading and communication technology; my early thinking centred on extracting private and sensitive information from people's minds for personal rewards, such as bank managers for the code to their safes. However, the truth was there were so many possibilities for its use, so much to consider, that while experiencing the technology myself, still questioning it and feeling tense because of it, I was incapable of thinking through any hypothesis.

We stopped along the way at various locations and at each stop I would get off the bus for refreshment, the toilet, for a cigarette or just to stretch my legs. At each stop the voices would tell me I was being followed and watched. I would look into the night sky to enjoy the stars, while also trying to identify the satellite again but the voices would instruct me not to do this and repeatedly told me to be very careful because someone was watching me.

As we neared Melbourne, Gita, the female voice, told me to leave the books I had bought on world religions on the bus. I questioned her instructions and asked why; her reply was, she had her reasons, and so I did what she told me to do.

By the time we arrived in Melbourne I had now

convinced myself that the hooded Asian man was working with the voices and after the coach had stopped tried to follow him or at least make myself easily approachable to him but he quickly disappeared into the night as we disembarked the coach.

Finding somewhere to stay near the coach station called Todd's Hall Backpackers, I booked in for two nights. It was now that Gita, the female voice, gave me further instructions to stay in the room for 48 hours, saying she wanted me to lay low for two days.

Then I experienced a whole barrage of unusual sounds that concerned and unsettled me, then listened as whispered conversations took place, the kind that you just can't quite hear, straining to hear what was being said and sensing that the conversations were about me. Knowing that it would be difficult to take 48 hours of these unusual sounds and whispered conversations I decided to disobey my instructions, leave the room and ring my sister, Sheila, in Sydney, to arrange a meeting time and place in two days' time. For whatever reason, my disobeying their instructions didn't cause conflict but rather seemed to gain the voices' respect.

Sheila didn't answer her phone and then Gita, the female voice, told me, "Sheila is not answering her phone because we've killed her," but by this time, after so many empty threats, I took little notice but became slightly more concerned after being unable to get hold of her after three or four attempts.

In the evening I made my way to Borders in South Yarra, as there was a book Gita had wanted me to buy by R. Otto. The evening was peaceful, time was spent reading Otto's book before having an early night. At one stage there sounded something like a gun shot outside in the street but I checked next morning and confirmed that it had only been a car backfiring.

The following morning I had breakfast in the courtyard of Todd's Hall then took a tram ride to St. Kilda's; it had been some weeks since I had seen the sea and it always drew me to it. I walked along the sea front and St. Kilda's Pier, looking at the starfish in the rock pools below before then returning to the comfort of a seafront restaurant to enjoy some fresh oysters and a glass of Chardonnay. At each stage of my day by the sea the voices told me I was being watched and followed. Unsure whether to believe them or not and trying to ignore them I nevertheless had an enjoyable day out by the sea.

The continuing suggestions that I was being followed and watched served to unsettled and unnerve me a little and on returning to Melbourne and because of running low on my supply of Temazepam, decided to visit a local chemist to get some more. Gita warned me that the tablets were highly addictive and that I should give serious thought to not using them, noting her advice but still buying them.

As I made my way back to Todd's I passed a university and stopped to read a poster supporting the withdrawal of Australian troops from Iraq. As I was reading it four young students walked by and one of them tore the poster off the wall. Challenging them as to why they had done that, the five of us got into conversation regarding the war in Iraq during which they invited me to join them for a beer in a nearby pub where we continued the conversation. We spoke for a further hour together before one of the students gifted me some marijuana as a present as we all said goodbye.

The following day I took the free tram ride to Docklands to see the Telstra Dome before setting out to discover

Melbourne itself. I walked along Swanston Street and visited the State Library and Melbourne Town Hall before heading towards Federation Square to visit the Ian Potter Centre.

Melbourne had a great feel to it, lively, friendly and exciting and as I walked Gita and I starting playing a game of trust together. I would walk up to the edge of the pavement and close my eyes, Gita would then tell me whether it was safe to cross or not, needing total trust to step out into the busy road and, finding it difficult in the beginning to take the first step, but enjoying the game more with practice.

The Ian Potter Centre was an excellent art gallery and there was one piece of art that really caught my attention. It was a wall approximately 50 foot long with just three paintings on it. The first painting was by Richard Grayson, it spelled out in ever decreasing letters the following statement:

"thecurrentinabilitytoconstructalternativetoglobalcapital ism"

The second picture, a lot smaller and in the middle of the three, was of a human skull; the third picture, bigger than the second, was of an atomic bomb exploding. It made me stop, look and think about what a strong communicator art can be.

From here it was back to Todd's Backpackers, playing and enjoying the trust game of crossing the road with Gita on the way. As we played together, she instructed me to not only ignore but also not to look at the posters that advertised a terrorist hot line.

The next morning I shared my breakfast with the birds in the courtyard of Todd's, relaxed and listened to the sounds of nature.

It was then I heard a new voice, it was a male voice, very calm, with a warm friendly tone. He encouraged me to relax, close my eyes and meditate. As I closed my eyes, a soft gentle rhythmic beat of a drum started in the background, the voice encouraged me to meditate on the sounds of nature.

He then repeatedly said, "Meditate on Allah, meditate on Allah," in soft slow gentle tones, with the soft slow gentle beat of the rhythmic drum in the background and the sounds of nature, the birds singing, the sound of the wind as it gently rustled the branches of the trees and in the warming sun I found myself drifting into a deep hypnotic or meditative state.

All of the above combined to produce a very powerful meditation, a peaceful, relaxing, beautiful state, allowing myself to enjoy it for a while but feeling that if I allowed myself to totally let go I would be out of control and I wasn't able to trust enough to do that.

It took me fully ten minutes to come round after the experience. This new male voice then asked me, "Why not do something special for your God, why not do something special for your God?" Because of all the events that I had experienced over the past weeks, I made a big note in my note pad. "The moral is – Do something special for your God."

After I had fully woken up I headed south over the Yarra River to the Botanic Gardens where, now feeling very relaxed and at ease, spent several hours in peace and tranquillity enjoying the excellent gardens, plants and wildlife.

In the afternoon I made my way to the National Gallery of Victoria to sample some of the art collections there, and then later in the day enjoyed tea and cake on the banks of the River Yarra, where I planned the final leg of my trip to Sydney, which would be later that night.

The voices now reminded me of what they were expecting me to do in Selfridges, Sydney and reminded me that I wouldn't leave Australia unless I did what they wanted. This made me nervous and, because of this, in my note pad under the statement, 'Do something special for your God', I wrote in capital letters, 'IN A PEACEFUL MANNER'.

Off to Sydney

It was to be another through the night trip from Melbourne to Sydney, taking about 12 hours; it would prove to be an eventful trip.

Positioning myself at the back of the coach and trying to sleep it was about half way into the journey that I started hearing Tottenham Man's voice again. It was threatening as he constantly harassed and insulted me; this went on for some time. It was also during this time that I started experiencing sensations of burning hot and ice cold again, as I had done in the RNLI club in Alice Springs. This time the sensations of hot and cold were aimed directly on a cut that I had on my foot which became painful. I looked out of the coach windows looking for the satellite and he said, "Yes we're still here, it's still us."

Memories of the nights in Alice Springs came flooding back; memories of the nights in Kashmir came flooding back, as I started to feel uncomfortable, started to get nervous and started to sweat.

Eventually I decided to try self-hypnosis in an attempt to sleep, hypnotising myself to the rhythm and sound of the bus; this worked to some degree as I think I had some sleep at some point during the night.

It was about five in the morning when I woke and starting reading my guide book on Sydney.

It was as I got to the pages on King's Cross, Gita told me in a serious tone that this was the real reason for my journey; all the synchronistic events that had happened to me along the way were about this. My journey was meant to be King's Cross, London to King's Cross, Sydney. She said I needed to pick up the sword of Jihad I had seen outside the hospital in Alice Springs, that they wanted me to be a suicide bomber, their first suicide bomber in Australia.

She went on to tell me that normally their suicide bombers were young and this had caused some people to question the bombers' motivation but I was mature, had a degree in Theology and would therefore have a significant effect. She also pointed out I would be famous worldwide.

The plan wasn't to put a bomb in Selfridges, Sydney but to get a suicide bomber to attack King's Cross, Sydney; it was an area representing all they were fighting against and that I was the perfect person to carry it out.

It felt at that stage as if something inside my mind had been triggered, as if some auto-suggestion had been placed inside my mind at some point, as I struggled and mentally fought to resist what they wanted me to do.

As we drove north towards the outskirts of Sydney, Tottenham Man spoke; he told me that he had a lot of friends in this area and that this wasn't some game and, if necessary, he would get someone to strap the bomb on me and throw me and it into the target.

Resisting them, it became clear to them that I wouldn't willingly do what they wanted me to do.

They told me this now caused a real problem, a real problem because I knew far, far too much, and that now I had become a liability, that I had become a problem that needed to be sorted out and needed to be sorted out quickly.

They went on to tell me that they would have someone meet me, have someone take care of me, as soon as the bus reached Sydney. For the remainder of the journey they keep on questioning me about whether I would change my mind but the more I resisted them the more they told me that I was going to get what I deserved.

As I disembarked the coach at the bus terminal by Central Station in Sydney an Asian man approached me with a camera in his hand and he asked me to take his photograph. I considered that this was the contact that the voices had threatened would appear.

Thinking that the camera might blow up in my face I held the camera up and it was with trembling hands that I pressed the button, uncertain of what would happen. To my great relief nothing did happen but the incident so shocked me that I was now shaking, visibly shaking from head to foot, because of what I thought might have happened.

Then in an effort to calm myself down I took a table in a nearby café and ordered some tea while waiting for my heart rate and pulse rate to slow down to a more normal level.

It was a massive moment for me!

After a few minutes and noting that there were two police cars by the bus stop, Gita told me that they needed to get me out of there as quickly and as quietly as possible. She instructed me to leave my rucksack in the left luggage department and get away from the station as quickly as possible. This I did.

Walking down George Street away from the station, with my heart and pulse rate now down but with high levels of adrenalin pumping through my body, she told me to find somewhere quiet to stop and sit, she also again told me to ignore and not to look at the posters that advertised a terrorist hot line. I was flying with the adrenalin rush and seeing all the tall buildings that lined the street; it was me who suggested we could bomb this, and we could bomb that. "No," Gita said, "we're not going to bomb anything, just at the moment."

Finding myself in Hyde Park I decided to rest there because at least the name was familiar, it reminded me of Hyde Park in London and therefore offered some kind of reassurance. Finding a tree and sitting under it Gita told me that I was now safe. I sat, rested and calmed down a little more.

Gita said that the situation now had to be reassessed, that it was a close call at the station and we needed to talk somewhere quiet and private; she told me to take a ferry ride around Sydney Harbour. This I did, walking on further to Sydney Harbour and taking one of the ferry tours. I sat at the rear of the boat as it sailed out past Sydney Harbour Bridge, past the Opera House, into the harbour itself. There, the sea air, slight breeze and sun helped to further calm me.

Gita now told me that I had come very close to being arrested at the station and that she needed to coach me on what to say if I was picked up by Australian Intelligence because I knew far, far too much. After about an hour of questioning and discussing the situation, the length of the tour, she told me that they were very uncomfortable and believed that if questioned I would tell all. She told me to make my way to the Botanic Gardens while they decided what to do about the situation. This I did.

Walking in the gardens for a while and seeing the spectacular views back towards the Harbour Bridge and The Opera House only helped to reduce the tension levels further. Some time passed before Gita told me to find somewhere to sit and to make sure it was in the open. This I did.

She then told me that they had decided that they had to kill me, it was the only thing to do, it was the only way out of this situation and that they would be sending an assassin to do the job. I sat and waited, reasoning that there was nowhere to run to; they had followed me from Kashmir, India, Australia and possibly England, around the world, there could be no escape. As I sat and waited, a small child came running over, chasing after his balloon. He stopped and smiled; I smiled back thinking it may be my last. I surveyed a man walking in the park, trying to identify the assassin. Gita told me, "No, it's not him but someone will be there soon." I sat, waited, calm, prepared to face my death.

Then Gita told me, "He's here now, you can have one last cigarette and as you put the cigarette out he will shoot." Putting out my cigarette, I said to myself, may God bless my soul.

But nothing, nothing happened.

A short time passed and then Gita said, "Fuck off, you're free to go, go on fuck off." Relieved and exhilarated, I thought to myself, back in the game, back in the game.

With that I started making my way back to the railway station, stopping to buy some water in a small shop on the way, where Gita told me someone was waiting for me but by now her threats meant nothing.

On the way back to the station I heard the male voice, the voice that I had heard in the garden of Todd's Backpackers in Melbourne, the one that had helped me to meditate, still in soft calm tones he said, "Fantastic, that was fantastic, come and join us, you must come and join us, we could use people like you." Flattered, I said I would consider it but at this precise moment felt too drained, too empty, too tired, to make any important decisions.

As I walked further towards the station I passed an internet café. Gita told me to go into the café. Disobeying her I refused and walked on, she then told me that I must learn to do what I'm told, to totally trust her because in an operational situation that was how it worked. Ignoring her I walked on to the station where I collected my rucksack and bought a train ticket to Riverstone, where I had arranged to meet Sheila at six o'clock that evening.

During the train journey Gita and I were in constant communication, she told me that I didn't have to be a suicide bomber if I didn't want to be one. That there were plenty of other things I could do for them, such as transporting explosives. With that she started coaching me in how to avoid arousing suspicion and the need to act naturally in an operational situation.

As we pulled into the station I noted some police cars out of the train window; Gita told me that they were there to arrest me and to stay calm.

I was about an hour early for meeting Sheila, so had a cigarette and paced about and explored a little of the area nearby, discovering that the police cars were parked at a police station next to the railway station in Riverstone and not there to arrest me at all. Gita told me she knew that there was a police station there and had just wanted to test me. I replied how could I trust her if she kept lying to me? I also discussed some of my theological ideas with Tottenham Man as the time slowly passed.

It must have been about six o'clock when Gita said to me in a very serious tone, "I need you to do something for me, pick up your bag." I questioned her, she said, "Just pick up your bag." I did. "OK," she said, "walk away from the station; just start walking slowly away from the station." Slowly I started walking away from the railway station, "Keep going," she encouraged, as I stopped to question what I was doing, "just keep going," she further encouraged, "just keep walking."

"Sheila is not coming," she informed me, "just keep walking, just keep going, Sheila is not coming, Sheila's dead, we have killed her, did you think we were joking, just keep walking, we're very serious, just keep walking, just keep going," she said in calm but authoritative tone; before I knew it I was halfway along Riverstone High Street.

"Further," she commanded, "just keep going, I will tell you when to stop." This must have gone on for almost an hour and by that time I was over a kilometre from the station.

"OK," she finally said, "here's fine, stay calm, stay very calm, I have something to tell you. We've kidnapped Michelle, she's well and safe and being looked after but there is something we want you to do for us, just do as you are told and everything will be alright.

"We've kidnapped Michelle because there is something we want you to do, just stay where you are and I will tell you what's going to happen. Just stay calm, just stay very calm.

"Wait where you are and a man will come and collect you and take you to a hotel there; he will tell you what it is that we want you to do. We weren't playing Andrew we are very, very serious. A car will pull up, you will get in that car, you will be taken to a hotel and you will do what you're told. Then your daughter will be fine she will be alright as long as you just do what you're told."

I waited for the car to arrive; as time passed and it started to get dark, I imagined what is was they wanted me to do. Maybe half an hour or an hour passed but with no real idea of time, with so many thoughts running through my mind, scrutinizing each car that drove by. Concerned for my daughter's safety I asked Gita when the car was coming. She eventually told me that there had been a change of plan and that I needed to walk back into the town centre.

When I had done that, she told me that the new plan was for me to get into a car that was parked nearby and drive it to a destination that I would be told later, she told me that the car had already been rigged with explosive and that the keys were on the driver's wheel but, she said, that they were going to kill Michelle, now the only question was whether they killed her quickly or killed her slowly. If I did what they wanted then it would be quickly, if not they would do it slowly.

It was at this stage that I heard Michelle's voice, it was really Michelle's voice and she said, "Don't do it daddy, I will be alright, don't let them make you do anything you don't want to do."

Michelle is a strong character and it would be the type of thing she would have said.

Gita then asked, "Do you want to speak to your daughter, then go to the phone box and pick up the phone," which I did, fully expecting to speak to Michelle but there was only a dialling tone.

Gita then told me, "There's another way out of this." She told me to walk down the road and she would tell me what it was. As I walked down the road, there was a large truck driving towards me, "Right you scum," she said, "throw yourself in front of the truck and then everyone will be happy, go on, end it right now." I considered it, for a moment.

"Too late," Gita said as the truck drove quickly past, "we've now killed Michelle."

Tearing off my rucksack and throwing it to the ground I shouted to the night sky, "You can't negotiate with Hiz-bella, you can't negotiate with Hiz-bella!"

Then I went into a nearby garage and asked them to ring

the police as I now believed members of my family had been killed by terrorists.

Tired of waiting for the police to arrive and becoming increasingly agitated I decided to go to the police station that I had seen next to the railway station. As I approached the police station, Gita spoke again, "We've now kidnapped your dad," then I heard my dad's voice, it was really my father's voice, "They've got me boy, got me tied to a chair, said they're going to kill me, don't give into them boy, don't give into the bastards."

It was my dad's voice, it was the type of words he would have used, and it was the type of fighting attitude he would have had. After his words the sound of a gunshot followed.

I told the desk officer that my family had been killed by terrorists, he asked me to take a seat and he would get someone to talk to me.

It was then that I heard Tottenham Man's voice again. "You've really done it now," he said, "three year bang-up in a nuthouse being raped all the time."

This was also the view held by the officer that came to deal with me. He handcuffed me, put me into the back of a police wagon, went to collect my rucksack which had been left on the ground by the garage and then took me in handcuffs to a local mental hospital.

During the journey to the hospital I could hear both the voices discussing what they needed to do. They said, "We've got to kill him now, got to kill him, we need one shot and that should do it, just one shot at him. We'll do it as he gets out of the truck at the hospital." Hearing this I resolved to make a quick dash from the police wagon into the hospital when we got there.

As we drew up outside the hospital I heard Lin's voice,

she said, "Andrew. What have you done? What have you done? There are Al-Qaeda terrorists in the house, they've shot my mother in front of me and now they're going to shoot me unless you give them a shot, whatever that is. Andrew please give them a shot, you know I have always cared for you, please give them a shot."

It was really Lin's voice and was the type of thing she would have said.

I didn't agree and with that Tottenham Man said, "No sorry Lin," laughed, and with that, there was another sound of a gunshot.

Sitting in shock as the doctor examined me, what was said or not is beyond recall. He left at some point. I was at a point of denial, at a point of blankness but totally believing that my family had been killed I sat in the interview room waiting for the doctor to return.

It was then, while sitting alone, that I felt a great surge of energy come into my body. Taking up all this energy I projected it mentally, with the full force of all the anger, pain and hurt that I was feeling towards Kashmir.

Being left alone only served to push me nearer to the edge, if indeed I hadn't already gone past that point. I went in search of the doctor or a nurse to provide some comfort and some reassurance.

I found a nurse, who made me comfortable in the main ward and provided me with a blanket; all the other patients had gone to bed. And it was there, shivering with fear, that I vowed revenge on the voices from Kashmir, not knowing how, not knowing when but swore revenge on them at some point, at some time for what they had done to my family.

It was about now that I started hallucinating, as Tottenham Man continued to taunt me.

After some time an ambulance crew came to take me to a different hospital. It was on the way to the other hospital that I heard the voices saying that it was a secure hospital; this made me feel safer, secure for me meant protection. As we arrived I heard the voice from the garden at Todd's Backpackers in Melbourne. He, still in a soft, calm tone, told me they had now kidnapped my other daughter Helen and for me to say nothing or they would kill her also.

The nurse who interviewed me seemed young and inexperienced. I told her that my family had all been kidnapped and killed by the terrorists. She asked me how I knew. Not wanting to tell her about the voices, I lied, I told her that I had received an SMS text message.

With that I was sectioned to a secure ward of Parramatta Mental Hospital.

The Following Morning

The first thing I did on waking was to go to the desk and ask the nurses there to ring the terrorist hotline and tell them that all my family had been killed by the terrorists. Disbelieving me they, instead of ringing the hot line, gave me an injection of some type to calm me down.

This must have helped as later that morning I rang Sheila. She answered and thank God she was alright and no harm had happened to any of my family. The rest of the day was spent reflecting and thinking about the events of the previous night.

But it was with utter and total disbelief that I watched that evening's news as it broadcast details of an earthquake in Kashmir. It was impossible to think that the events of the previous night and the earthquake in Kashmir were related in any way, yet given the journey I was on and all the experiences that I had had, there was a link in my mind.

That is the reason for the dedication of this book:

For the people of Kashmir;
May their prayers be answered and
May their leaders serve them well.

The Next Few Days

The next day Sheila came to visit. I was feeling a lot calmer, although slightly psychologically sore after the experiences of the previous two days and after she had gone I spent time relaxing and talking to the other patients.

The day after, I saw the psychiatrist. For this meeting there was only one, with his assistant, and because I had lied about the SMS text message I decided not to go into any details regarding the voices but did discuss the panel of ten psychiatrists and doctors that I had seen in Alice Springs Hospital.

He expressed great surprise that the hospital, although a lot smaller than Parramatta, should have such a resource. The relevance and importance of that meeting in Alice Springs, where the panel of ten psychiatrists and doctors had asked if I was happy to continue on my journey with the voices, to see where it took me, still did not register.

Because of the type of injection they had given me two days ago, he said he would like to keep me in for two more days for observation but prescribed no other medication.

The next two days were spent relaxing and resting, and going through my note pad. It was while doing this one day that another remarkable event happened.

The wind caught the page of my note pad, on which I had written, 'The moral is – Do something special for your God' and then at the Yarra River in Melbourne, added under it, in capital letters, 'IN A PEACEFUL MANNER'. This caused the page to flip over and creased it. But the crease on the page was exactly under the words, 'IN A PEACEFUL MANNER', as if to highlight that point.

The crease was so close to the actual wording that Gita tried to argue that the crease was meant to go through the wording and cross out the words, thus invaliding the statement and not to highlight the statement. It was a point we would revisit later.

Sheila came to visit both days and on the second day she took me to her home in Maroota.

The Next Month

Sheila lived in Maroota, in a gorgeous ranch style house set on 40 acres of bush, about 30 kilometres outside of Sydney with her husband Steve and their two dogs; I was to spend the next month enjoying their company and hospitality while recovering, reading, thinking and researching some more.

Our first day out was to Darling Harbour, taking the Rivercat from Parramatta to Sydney and while there enjoying a fabulous meal of freshly caught barramundi washed down with some wonderful Australian Chardonnay. The love that danced between my sister and I helped to settle and secure me.

At night I told her of the satellite and how it had been reading my thoughts and how it was still doing so. She said that although she could see I was mentally well that she found the story of the satellite a bit like science fiction and therefore difficult to believe; she also found the whole thing frightening. Understanding this, I didn't go into too much detail, in any event we had a whole lot of conversations to catch up on, having seen each other only twice in the past 15 years.

As we spoke a car drew up in the distance at the top of the drive, its headlights shining brightly towards the house, clearly visible in the darkness of the bush. Gita, the female voice, told me, "That's our car." I didn't believe her.

"Right," she said, "I'll prove it, I'll get the headlights to flash." With that the headlights switched off then back on again, still disbelieving her, putting it down to just coincidence. "Alright," she said, "let's do it again." With that, sure enough, the lights flashed off and then back on again.

This slightly unsettled me but I didn't tell Sheila. If the voices had wanted to do any physical harm, there was ample opportunity in the outback, so why do it now? They were just letting me know they were still around.

The weekend was excellent. Time was spent resting and playing with Sheila, Steve and their dogs. Nicholas, Sheila's son, my nephew, came over for dinner and we all enjoyed the whole family atmosphere. I asked Sheila if there was a Selfridges in Sydney, not telling her the reason for my question; she told me that there wasn't one, which came as something of a relief.

A small cyst that had been on my head for over two years

disappeared and because of this I recalled a conversation that I had with Ashley while in Kashmir, about how the Sufi Baba had prayed for one of Ashley's friends which resulted in a small mark disappearing and I believed the Sufi Baba, in Kashmir, had said a prayer for me and had helped the cyst disappear.

Sheila had a spare 4x4 that she let me use for the duration of my stay, which enabled me to go exploring locally. The first thing I wanted to do was visit the Baha'i Temple because I had seen their stall in Mindil Beach Market in Darwin and their temple in Delhi. Secondly, I wanted to visit Bali el Huda Mosque because Richard Khan, the opal mining poet from Coober Pedy, had invited me to visit his mosque.

The Baha'i Temple was a lot smaller than the temple in Delhi and although the grounds were lovely the temple itself didn't really evoke anything within me.

The mosque on the other hand was quite different, the imam invited me into his living room to discuss faith and Islam, gifted me a book called *Revelation, Rationality, Knowledge and Truth* before inviting me to join him in his worship.

Finally, I wanted his advice, telling him that I had been following my path for some while and was now in need of some guidance, without going into any details regarding my experiences.

"Ah," he said, "we in Islam recognise this well," he said with confidence, "pray more, pray more however you pray, as a Christian or as a Jew or as a Hindu or however you pray, pray more." Smiling, I left and returned to Sheila's house in Maroota.

Sheila had taken our first week together as holiday and

we were therefore able to enjoy being with each other as she showed me around. We went to Hawkesbury River, Wiseman Ferry, some local vineyards to sample and buy the local wines. She also took me to the crematorium were my mum had been cremated. I would also spend time reading the book that the imam for Bali el Huda Mosque had gifted me, playing with the dogs and enjoying the sights and sounds of the bush.

At night the sky was peppered with stars; there was little if any artificial light around and therefore they shone as brightly as they had done in the desert. We would use binoculars through which you could clearly see the Milky Way, and some evenings there would be thunderstorms, the power and clarity of which I had never experienced before.

The second weekend I wanted to visit Hillsong Church, an evangelical branch of the Methodist Church, on the outskirts of Sydney. It would be the first time I had been out at night on my own since my experiences in Riverstone. As I drove through the bush, Tottenham Man threatened to blow out the tyres of the 4x4, with the laser that had stunned me in Alice Springs and cause me to crash. Ignoring him I drove on.

The church was packed with around a thousand people inside. On stage there was a ten-piece band and a choir of maybe 20 people to one side. The atmosphere was electric with people standing, singing, clapping and praying. At one stage there was a prayer for a particular member of the congregation, who had recently lost a close friend. This only added to the already highly charged emotions within the church.

After the service I left feeling uncertain of what to make of this evangelical type of worship. Gita gave me her views, saying, "That it was well delivered and professional but a bit showbizzy," which I had to agree with. It was disappointing to learn that the church is facing allegations regarding their raising and use of tithes, a big money business, the God business.

The next day we all went to visit Nicholas to see his new house and meet his girlfriend's family, which was a good day but when we returned the dogs had gone missing. Gita now told me that they had taken the dogs but by now the threats concerned me less and less. The dogs returned the next day, after one night away, having just gone off into the bush for a night.

The following week Sheila returned to work and I used the opportunity to do some work on her computer.

Firstly I wanted to learn what was happening with regard to the earthquake in Kashmir. I was appalled to discover that India and Pakistan were bickering over the rescue effort there. It appeared that India had declared that they could cope and didn't need any international help, while Pakistan was asking for help. India had indicated that they could loan some helicopters to Pakistan but only on the condition that their own pilots flew the helicopters. Pakistan were unwilling to accept this help with these conditions.

The only people to suffer in this stand-off were the people of Kashmir who had lost tens of thousands of their countrymen in the earthquake, with hundreds of thousands more made homeless. This state of affairs so concerned me that I wrote to both governments expressing my views.

The second area I wanted to do some work on was researching the subject of telepathy, discovering that something called the Ganzfeld Method had been used in research into telepathy. This method of research included, amongst other things, the use of white noise in the experiments. This instantly reminded me of the radio I had used on the houseboat in Srinagar, Kashmir, in an attempt to block out the voices but could only find white noise, lots and lots of white noise and considered that if by using the radio I had assisted the voices in the reading of my thoughts.

Also I discovered that research into telepathy dates back over 40 years and the public in general have a high level of interest and belief in the subject.

Further, I found out that each person's brainwaves are individual, much like a fingerprint. Maybe it was by identifying these individual brainwave patterns that technology could now intercept and interpret the small electrical charge given off by the brain, thus being able to read thoughts, remote telepathy. These discoveries kept me busy for days thinking about the subject.

It was also during this week that Gita encouraged me to fast for a day, as it was the last week of Ramadan. This I did, finding the experience very bonding, knowing that up to one billon other people were doing the same thing as me at the same time.

Also, as it was the last Friday of Ramadan, I decided to visit the mosque in nearby Annangrove. I had heard that the imam had experienced a great many problems as the mosque was being built and I felt that I wanted to show him some support on this special day.

Leaving Sheila's at about eleven thirty in the morning, with Tottenham Man demanding, "Get a move on or you'll be late," I stopped for petrol on the way, which caused him to further encourage me to hurry up. I got there at exactly midday, coincidently the same time as somebody called Fazlul Huq. Fazlul introduced himself to me, saying he was a doctor and a poet, and then introduced his latest poem to me, it started, "The life is a rendezvous with fate, As the nafs is a traveller through the gate!"

Nafs here means soul. It reminded me of something Ashley had said to me in Kashmir which was, "when he is born so it is written."

With that we made our way into the mosque. The imam had delayed the prayer for some moments, waiting to see if anyone else would arrive. One of the congregations turned to us and said, "We were waiting for you."

The prayers focused on the relationship between Allah and the individual and how in public there was a need for certain rules and customs to be observed but how that it was in private that the individual and Allah could really talk, really be close. It was the second part of the prayers, the individual's private relationship with Allah, which I strongly identify with.

After the prayer Fazlul and I left together; he gifted me a book of his poems as we said goodbye.

As I drove down the road the voices started with, "Ooooooooo-d-i-ooooo you're a Muslim now, you're a Muslim now."

Of course I wasn't but this time the sound of Ooooo-d-i-oooo didn't frighten or concern me, in fact I rather enjoyed it.

That weekend on the Saturday Steve and I went for a long walk in the bush during the day and in the evening watched a film called *Enemy of the State*, which deals with how technology can now be used to track individuals, technology that wasn't even thought of ten years ago. One of the first lines in the film is "Inside your head is now the only place that you can be private." This made me smile because I now knew this was no longer the case.

On Sunday Sheila took me to see the home where my mother had died, which was emotional.

The following week was to be my last with Sheila and Steve as I planned to travel back to Brisbane the next Sunday. It was spent reading, researching and thinking, playing with the dogs and training them to return sticks and balls during the day and enjoying eating, drinking, relaxing, talking and being with Sheila and Steve during the evenings. We celebrated Australia Day with some of their friends in typical Australian manner, barbeque and beers, and had a day visiting the Blue Mountains which are really blue; our final day out together was spent locally.

We visited the nearby Ebenezer Church which is the oldest operating church in Australia, this was of great interest. Opposite the church was a tree, under which services were held before the building of the church in 1809. The graves were mainly of Europeans with only a very small corner of the graveyard allocated to the indigenous Aboriginal.

The next day Sheila took me to Sydney Airport, we said goodbye and I was off to Brisbane.

Return to Brisbane

Sharon met me at the airport and took me first for dinner at a local marina, where we met some of her new friends and then afterwards to a party, which went on through the night. This was where I discovered Midnight Oil, an Australian rock group, whose songs and lyrics reflected and echoed many of the themes and thoughts that I had had regarding the indigenous Aboriginals; their singer Peter Garrett is now Australia's Environmental Minister.

It was in the early hours of the morning when we returned to her house in Beenleigh and she told me that she and Aussie were going ahead with their separation, although still living together at present. She invited me to stay with them for the next couple of months during this period of transition, which I was more than happy to do.

The first thing that I needed to do was to renew my visa as it was about to expire, planning initially to do the visa run to New Zealand or Fiji before discovering that it could be renewed online from inside Australia for a small fee.

During that first week back in Beenleigh I told Sharon and Aussie of some of my experiences while travelling around Australia but could tell that they made Sharon feel uncomfortable and apprehensive, therefore I didn't go into too much detail. However, she did raise an interesting question: Why you, why would anyone choose you, to do that to?

This was also a big question for me and one that I had asked myself many times during my travels. It was during that first week back with Sharon, while researching on the internet and with huge relief that I found a website called Mind Justice, which highlighted accounts from other people who were going through similar experiences to me.

I cannot express enough how reassuring it was to find out that I was not alone in hearing these voices and having these experiences. The website gave details of how the technology worked along with addresses of other websites that gave further details.

During this period Aussie and I again kept each other company and played the Pokies at the nearby Beenleigh RNLI club in the evenings, while my days continued much the same as they had at Sheila's, reading, thinking and researching on the internet with Gita and Tottenham Man, in daily if not constant contact.

I was still hungry for news from the earthquake in Kashmir and it was with tears rolling down my checks that I read the following quote: "The earthquake felt like the day of judgement. I thought it was the last day ever. But then, after some time, I remembered that according to Islam Judgement Day is on a Friday. The quake happened on a Saturday so it couldn't have been Judgement Day."

This expressed such a strongly held belief, such a strongly held faith that combined with my own experience on the night of the earthquake made me want to revisit the Koran.

My second weekend back in Brisbane was spent going to see my nephew Zane in a school concert on the Saturday. And on the Sunday one of Sharon's friends called Trina had offered to take me out for the day fishing with her parents. Her father was a retired ferry captain so knew the local waters well. We set off in his superb wooden boat across Moreton Bay to Stradbroke Island or Staddie, where we caught some bait before then heading out into the bay once more to do some fishing proper. Then cooking some of the freshly caught fish, it was an excellent day out.

The next week time was spent reading a book on Padre Pio; it was a book I had bought along my travels because, when seeing it, it reminded me of the service that I had attended at The Church of Padre Pio in Victoria, London some months back. It surprised me to learn that "The life of this man was characterized by some of the phenomena typically associated with the paranormal; bilocation, levitation, mind reading..."

The book also discussed aspects of his life that matched closely with the experiences that the retired Bishop of Gibraltar had told me about, also some months ago, back in Gibraltar.

It was while sat reading this book on Sharon's deck that I noticed a small stone in the shape of a face. As I looked closer, it became clear than it could be seen as a Hindu woman's face, Hindu because the bindi in the centre of her forehead was clear to see and therefore identify her as such. When Aussie returned home from work that night I showed it to him and asked if he could also see the Hindu face in the stone, which he could. From that day on I would look at the Hindu face in the stone each day and smile inwardly.

It was also around this time that I found a website which detailed work being done by the Nobel Prize winning scientist Brian Josephson, into the field of the paranormal at the University of Cambridge. He was also working to reveal scientific discoveries that the government wanted to keep secret at present.

Sometimes Tottenham Man, the male voice, would issue threats but I reasoned that if they had wanted to do anything then Sheila's house or the outback would have been a better place for them to do anything. Occasionally his threats would unsettle me but in the main they had little effect. Gita, the female voice, and I would communicate continuously but this communication became increasingly friendly.

Sharon bought me a copy of the Koran for my birthday which I starting reading on her deck. Each time before I picked it up Gita would tell me to wash my hands and each time I put it down she would ensure that I didn't put anything on top of it.

As I started rereading the Koran, for the first time I could see the beauty within it that people spoke of.

One day, while reading a passage about how a spider had built a web across the entrance to a cave to provide camouflage, when as I looked up I saw a large spider building a web in the corner of the deck. Looking down at what I had just been reading and then looking back to the spider, a direction that I had looked in many times in the past but had never seen a spider there before and I strongly believed that this coincidence meant that what I was reading was the truth.

Gita told me that I was a natural Muslim and then started to encourage me to think about becoming a suicide bomber. It was a proposition I would give serious consideration to over the next weeks, maybe after all it was via use of this technology that suicide bombers were recruited.

Later that week Aussie and I were having one of our evenings out at Beenleigh RNLI club when an enormous thunder storm began. The storm was wild, with winds so strong that the trees bent at 60 degree angles, the lightening lit up the sky like daylight, the thunder was extremely loud, rain lashed into the doorways, it was the strongest storm I had ever experienced. A thunderbolt hit the club, knocking out all the electrics, part of the ceiling in the restaurant area collapsed as water poured in. Gita told me that the storm was because God was angry with me for drinking and gambling. I didn't agree but thought that the storm must have some meaning. It was on our next visit that we found out that on the evening of the storm someone in the club had won four and a half million dollars on keno!

My nights' sleep now started to be disturbed again. In addition to the voices, a new dimension was added, this was in the form of small electrical impulses that would cause my body to react in a spontaneous physiological manner, almost like a short sharp electric shock that caused a reaction, and sometimes there was a zapping to my eardrums which was painful. I kept these experiences to myself, feeling able to cope with them.

Some weekends Trina would come to visit and we would have days out together to Tamborine Mountain for lunch, a local spiritualist church and to Jupiter's Casino on the Gold Coast. One of the best was a day in Byron Bay, a hippy hangover, where we enjoyed the beach, the sea with its strong riffs and the market with its drummers, smoking drum circle and its spirituality stalls.

In was also during this period that I started writing about my experiences. Zane and Rhiannon had their school holidays and would keep me company, their friends would come to visit and we would watch films. One I recall was *Turner and Hooch* where one of the characters wears an amarillo shell on his head because it protects him from the CIA who he believes are reading his thoughts.

It was after much thought that I confirmed to Gita that I didn't want to be a suicide bomber. There were four main reasons for this, on the list that I made, apart from any moral considerations.

Firstly, there was nothing in all that I had read that encourages this course of action. Secondly, it wasn't in my nature. Thirdly, it was directly after seeing the sword outside Alice Springs Hospital that I had learnt about the inner and outer Jihad. All the signs, all the coincidences that I had experienced guided me towards the inner struggle. All in addition to the main reason, which was the page of my note pad that had been blown over by the wind in Parramatta Mental Hospital and creased the page, thus underlining the words: IN A PEACEFUL MANNER.

Gita tried to argue that the crease caused by the wind went though the words, thus invalidating them. Therefore, I cut along the crease to confirm that it underlined and highlighted the words, which it did, absolutely, completely, remarkably.

The process of considering becoming a suicide bomber caused me to redefine my views on terrorism from those that I had expressed to Ashley on the houseboat in Kashmir, now believing that a terrorist can be defined as someone who intentionally targets innocent people with intention to do them harm. Further, that the use of the word martyr to describe suicide bombers now appeared totally inappropriate; a martyr is someone who is prepared to die for what they believe in. Whereas somebody who intentionally targets and kills innocent people is a murderer!

Christmas came, Sharon, Aussie, Zane, Rhiannon and I enjoyed it with friends of theirs. Early in the New Year Aussie took the children on holiday to Noosa and on returning he and Sharon went their separate ways. As he moved out he said possibly the nicest thing anybody had ever said to me, "The angels brought you here to be with me at this time." He had handled the separation very well; although obviously a very painful experience for him, he behaved with honour, integrity and love throughout the whole event.

For me it was also time to leave and return to England

but first I wanted to fly back to Sydney to say goodbye to Sheila.

Sydney is the most beautiful airport that I had ever flown into. As the plane circled above the airport you could see the many inlets below with numerous boats sailing on them, as people played and enjoyed the sun, and then flying out over the beaches and the whirling sea that pounded into shore, before landing.

Sheila met me at the airport. The first thing she said to me was, "You'll never believe it but the tree outside the church in Ebenezer has been destroyed by lightning." It was the church we had visited on our last day out together; we made plans to revisit it the next day.

That's what we did the following day and, sure enough, the tree that had stood there for over 200 years outside Australia's oldest operating church had been destroyed by lightning. We took photographs and I tried to understand the meaning and coincidence of it. From here we went on to visit Tizzana Winery that markets itself as 'A touch of Tuscany on the banks of the Hawkesbury River', where we enjoyed lunch, wine and conversation.

Our week together passed all too quickly and it was time to return to Brisbane, say goodbye to Sharon, Aussie, Zane and Rhiannon and catch the plane back to England. It seemed almost surreal as the plane flew along the east coast of Australia; the events, the sights and experiences that I had experienced had been incredible.

During the flight Gita wanted me to confirm that I had agreed to give up the rights to any intellectual properties which may have come out of the journey. Having already agreed to this, I readily did so again.

It had been an amazing trip!

Return to England

Helen, my youngest daughter, met me at the airport; she had flown back from Canada because her Grandmother Pat had been taken ill and was in hospital. She took me back to Bushey and the four of us, Lin, Michelle, Helen and I, spent a few days together, the first time the four of us had been together as a family on our own for any period of time since the divorce over 16 years ago.

I told them of some of my experiences whilst in Australia but as with Sheila and Sharon I could tell that the stories made them feel uneasy, particularly in view of the condition I returned from India in and therefore again didn't go into much detail. After a few days Helen headed back to Whistler in Canada, I stayed with Lin and Michelle for a further two weeks; sometimes during those evenings I would tell Lin some more details of my trip, she was always very supportive and open minded regarding the experiences.

It was also during those two weeks that I decided to return to Gibraltar, not because I had any plans but because I wanted to complete my trip and by returning to Gibraltar I was doing that, but I wanted to visit my dad in Thame and my friends Gill and Stuart in Oxford before leaving.

It was during my trip to Oxford that Gill said something strange, she said, "I've got to tell you, The Alchemist." She went on to say it was an answer to a question she had seen on TV and didn't know why but knew that she had to remember it and to say it to me for whatever reason. I told her of the girl I had seen reading it at the Backpackers in Alice Springs and how I had recently read it along my travels.

I went into the town centre to buy it for her as a present

while also buying *The Non-Local Universe* by Nadeau and Kafatos, a book about a new kind of physics, thinking it may offer some insights or some answers to all the experiences that I had had.

With all my goodbyes said it was back to Gibraltar as Tottenham Man taunted me with "What's he going to do when the money runs out, what's he going to do when the money runs out?" while also threatening to have me kidnapped in Gibraltar and then taken across the straits to nearby Morocco.

His threats had little if any effect but he was right about the financial situation, my gap year was coming to an end and it had naturally drained my financial resources and I returned to Gibraltar with about 30 pounds left in my pocket after some unbelievable, incredible life changing experiences.

My friends Peter and Annie met me at the airport and I felt eager to tell them of some of the experiences that I had along the way but of course what do you say to a friend who is telling you that they are hearing voices and that those voices have been following them around the world and that those voices are coming via satellite. Sounds a bit questionable!

But at least Peter was very open minded and agreed that thoughts are only energy. We also discussed how technology is advancing at an incredible speed and what we now take for granted was un-thought of in the not too distant past. Later that night I met Jenny and we enjoyed Champagne on her balcony till sunrise and I gave her the wall tapestry that I had bought at Bernard's house in Srinagar.

The rest of the week was spent catching up with friends, in particular Jenny and Gordon, visiting old haunts and trying to decide what to do next or discover in which direction I would be taken.

This was when Lin rang to say she was coming to Spain for a holiday. We spent about a week together during which time we decided that we would both travel back to England and as we were both at crossroads in our lives stand shoulder to shoulder with one another.

My gap year was over but my journey went on.

The Understanding

The Next Eighteen Months

We returned to England and before long found a wonderful rose-covered cottage to live in; Michelle came to live with us. Then I found a really suitable job working for a sales agency called CPM on behalf of Procter and Gamble and life was pretty idealistic. We would sit in the secluded walled garden, drinking wine, watching the foxes, squirrels and birds come to feed and play, friends came to visit, Helen came back from Canada for a while. The whole family life situation was something I really loved; it was a very happy time. Incredibly, after following all my coincidences, after following all the synchronicities it had brought me here, to a happy idealistic family life, something that I had lost many years previously.

The voices were still with me and in the early days they used to zap my eardrums most nights which was painful and, because of this, I needed to put cotton wool in my ears for protection. Gita, the female voice, was in constant contact but Tottenham Man, the male voice less and less so, which not only meant that there were little if any threats but also that I became increasingly confident in communications with Gita.

Via Mind Justice, the website that I had found while at Sharon's house in Beenleigh, Australia, I was in contact with other people in England who were experiencing similar types of symptoms as myself and arranged to attend a meeting with them in London.

The venue was the Adidas gym, above Waitrose, in Canary Wharf; it was with a mixture of trepidation and excitement that I travelled there. The meeting was attended by about ten people with a range of ages and nationalities; it was clear from what was being said that they were indeed having similar experiences as me. They each had their own stories and each had their own interpretations as to what was happening and who was most likely responsible for it.

We arranged to meet in about a month's time; they would become known as my crazy friends to the family because there, sat round a table in the gym, above Waitrose, in Canary Wharf, London were ten adults having conversations about hearing voices that would not have been out of place in any mental hospital.

After meeting with them for the first time I considered it likely that it was not a terrorist organisation that had been intercepting and interpreting my thoughts but more likely a government agency because the technology involved would be extremely complex and therefore very expensive but I couldn't be certain.

It was also at this time that I reconsidered my experiences in Alice Springs Hospital when the panel of ten psychiatrists and doctors had asked, "Are you happy to continue with the voices, to see where it takes you?" This I now realized would of course have meant that they must have been aware of what was happening and therefore at least aware of whom was responsible, if not part of it.

The happy home life went on, Michelle teaching, Lin working for the police and me driving to visit customers doing my sales job and communicating with Gita as I did so.

Then one day Michelle decided she needed to go to the Science Museum, London to do some research for a forthcoming school trip; she invited Lin and me to go along for a day out.

It was with wide-eyed amazement that I came across an

exhibition in the Antenna Gallery; the exhibition featured work being done in the field of Neurorobotics, the frontier between the mind and machines. The exhibition charted the development of this new technology through its use in artificial limbs to the present day. It was supported by hands-on experiments that invited children to compete to see who could move a ball the furthest by the use of the mind, in addition to surveys that asked ethical questions such as: should science be allowed to read people's thoughts? There was also a display about how science was now able to zap your brain to stimulate particular parts of it, which would in turn enhance the abilities of those areas that had been stimulated.

This was the London Science Museum, one of the foremost in the world, confirming what my crazy friends and I had been experiencing; it was at this point that I knew that I had to start writing this book. I texted my crazy friends and it was at the Science Museum that we held our next meeting.

It was also at this time that I started to read Nadeau and Kafato's book *The Non-Local Universe*, but the book was outside my education and therefore I didn't understand the physics involved. However, I did understand that they were claiming 'the most momentous discovery in the history of science' and this claim was based on the Quantum Theory of Entanglement, which I had read about on my journey from Oxford to Cambridge. Also I did understand their conclusion that "the connection between mind and nature is far more intimate than we previously dared to imagine."

This conclusion not only supported the world view held by the indigenous Aborigines but also the view offered in The Gospel of Thomas, "That the Kingdom of Heaven is spread upon the earth but men do not see it." Their work has undoubtedly contributed to the field of Neurorobotics and the ability of science to now intercept and interpret the energy of thought.

To reconfirm the point being made, science is able to use technology to interpret the energy of thought.

It would seem that the technology was originally developed at close quarters but over time it has been developed so that it can be used over ever increasing distances and, as space is only something like 50 miles away, this now means it can be used via satellite.

It was from this point that Gita and I started discussing the implications of these discoveries.

Firstly, it seemed clear that the authorities, whoever they may be, now wish for these scientific discoveries to be in the public domain. This to my mind must be the case otherwise why allow the exhibition at the London Science Museum to take place or this book to be written?

This technology could offer enormous benefits, for example in the field of anti-terrorism. Terrorism is very real, many of the experiences that I have encountered during the course of my journey have been merely a psycho drama, very real and at times very frightening to me but nevertheless a psycho drama. In areas such as Kashmir the threats would be real and would have been normal occurrences at the height of the troubles and many, many thousands of innocent people have been killed.

This is cutting edge technology in every sense of the word and is I strongly believe currently being used in the field of anti-terrorism. Its use would explain how London and Glasgow avoided three bomb blasts on one day in 2007. We the public were told that luckily three bomb blasts were avoided by: One, a traffic warden who had a truck towed away to a safe place underground because it was illegally parked. Two, an eagle-eyed ambulance driver spotted some smoke coming from another car that was also made safe. And three, a suicide bomber got his figures wrong and was unable to drive into Glasgow Airport foyer and blow it up as he had planned to do. All of these things may be the case but to my mind it is more likely that remote telepathy was used to ensure the public's safety.

This technology could also have vast implications in areas as diverse as interrogation and psychology, there could be enormous commercial implications, it could revolutionise the way we communicate. There would additionally naturally be many ethical questions regarding its use.

But the main point for me regarding the discovery is what it says about what it is to be a human being. What is says about our natural state. What it says about theology, what it says about the science of God.

The idea of thoughts existing beyond the mind is something that finds expression in many of the great world religions, for example both the Holy books of Islam and Christianity talk of God knowing our every thought; now it appears that science is able to confirm this point or at least can confirm that thoughts can exist outside of the mind. As already discussed within this book, the idea, belief in and use of telepathy finds expression not only in the yogi of Tibet but also the indigenous Aboriginal of Australia, while the idea of our consciousness being entwined with others or with a greater consciousnesses finds expression within Hinduism.

It could lead to a new discourse between science and religion, Science, that so successfully marginalized religion in the West during the twentieth century, now wanting to address these new discoveries as part of and in light of ancient writing.

These were some of the topics Gita and I discussed as I drove around the M25, the motorway that encircles London.

Then one day Gita gave me a deadline to complete this book by; situation, circumstance and events all combined, meaning that the only way I could reasonably do it was to give up my job. So that's what I did.

Maybe that was what Marian had meant when he said, "Write, you first," as we said goodbye at Gibraltar Airport, all that time ago.

And, so how to end this book a coincidence is fitting. One day, while I was writing up some notes on Sharon's deck, about how the laser had stunned me in Alice Springs, when Rhiannon, my niece, and Zane, my nephew, came over playing with a laser toy, which they had just bought and with the small laser dot shining on the ceiling said, "I know how you can end your book Uncle Andrew; you could say I went on a journey and wrote about it, why don't you go on a journey and write about it too."

It is true, God is the Greatest.

Allah Akbar.

ABOUT THE AUTHOR

Andrew Cole has spent the majority of his working life in sales and sales management, in the FMCG sector. Working for a range of multi-national companies both directly and indirectly, these include: Rowntree Mackintosh, Timex, Seagram, Proctor and Gamble and for the last ten years McCurrach on behalf of Pepsico. He retired two years ago.

Like many he has experienced the vacillations of life and his own particular salvation was graduating at the Sheldonian Theatre, Oxford from Westminster College, in the early 1990s after completing a mid-life degree in Theology.

From here he spent ten years in the bookmaking industry, most of which was in Gibraltar working for a progressive bookmaking company as Head of Horse Racing Trading. It is the period after leaving that role that this story relates to.

He has two daughters both teachers in their own specialist fields and two young grandchildren, all living in Queensland, Australia, while he and his life partner Lin live a happy retired life in Oxfordshire, England.

PRINTED AND BOUND BY:

Copytech (UK) Limited trading as Printondemand-worldwide,
9 Culley Court, Bakewell Road, Orton Southgate.
Peterborough, PE2 6XD, United Kingdom.